Chemical Ionization Mass Spectrometry

Author

Alex. G. Harrison
Professor of Chemistry
Department of Chemistry
University of Toronto,
Toronto, Ontario, Canada

CRC Press, Inc.
Boca Raton, Florida

Library of Congress Cataloging in Publication Data

Harrison, Alexander G.
  Includes bibliographical references and index.
  1. Chemical ionization mass spectrometry. I. Title.
QD96.M3H37   1983   543′0873   82-14640
ISBN 0-8493-5616-4

PREFACE

The technique of ionizing a sample of molecules by gas-phase ion/molecule reactions was first reported in 1966 as an outgrowth of fundamental studies in gas-phase ion chemistry. In this pioneering paper, Munson and Field suggested that the method, which they called chemical ionization, might have useful analytical applications. The past 16 years have proven this prophecy to be abundantly correct. Chemical ionization provides information which is complementary to electron impact ionization in many cases and the two methods together are particularly powerful. As a result, chemical ionization is one of the most widely used ionization methods in mass spectrometry and it has found extensive application in structural elucidation studies and quantitative analytical studies in many branches of chemistry and biochemistry and in medical and environmental areas. At the same time chemical ionization studies have provided a large body of information concerning gas-phase ion chemistry.

The literature of chemical ionization mass spectrometry is now enormous and ranges from basic studies of the ionization processes to applied analytical problems where chemical ionization was the ionization method of choice. Numerous review articles have appeared covering various aspects of this literature, but these have necessarily been limited in scope. It appeared that there was a need for a more complete discussion of the field, particularly one which brought together the basic work which forms the body of knowledge upon which the applied studies are based. This is especially true because many people beginning work with chemical ionization mass spectrometry do not have a background in gas-phase ion chemistry. Accordingly, in the present book, I have chosen to emphasize such basic studies at the expense of the applied studies. In particular, the extensive fundamental studies of ion/molecule reaction kinetics and equilibria form the foundation upon which chemical ionization mass spectrometry is built and I have considered it desirable to review this work in some detail. At the same time I have not attempted to include the basic principles of mass spectrometry; these have received excellent coverage in other books.

The time necessary to write the manuscript was made available by a sabbatical leave for which I am indebted to the University of Toronto. The last half of the leave was spent at the École Polytechnique Fédérale, Lausanne, Switzerland, and I am indebted to the Institut de Chimie Physique and to Professor T. Gäumann for their kind hospitality. The excellent typing of Suzanne McClelland was invaluable as was the drafting work of Frank Safian. I am particularly indebted to the members of my research group and to my wife and family for their forbearance during my labors.

A.G.H.
Lausanne, June 1982

THE AUTHOR

Alex. G. Harrison, Ph.D., is Professor of Chemistry at the University of Toronto.

He received his Ph.D. in chemistry from McMaster University in 1956. After post-doctoral work at McMaster and the National Research Council of Canada, Ottawa, he joined the staff of the Department of Chemistry, University of Toronto, as a Lecturer in 1959, becoming Professor of Chemistry in 1967. In 1975, he spent six months as Visiting Professor in the Department of Molecular Sciences, University of Warwick and in 1982 he spent six months as Visiting Professor in the Institut de Chimie Physique, École Polytechnique Fédérale, Lausanne.

Dr. Harrison is a member of the Chemical Institute of Canada, the American Chemical Society, the Royal Society of Chemistry, and the American Society for Mass Spectrometry. He has served on the Board of Directors of the latter society and as chairman of the Physical Chemistry Division and Member of Council of the Chemical Institute of Canada. He is on the Editorial Advisory Boards of *Organic Mass Spectrometry* and *Mass Spectrometry Reviews*.

Dr. Harrison has authored or co-authored more than 150 papers in the area of mass spectrometry and gas-phase ion chemistry. His research interests lie in the chemistry of gas-phase ions and analytical applications of mass spectrometry.

TABLE OF CONTENTS

Chapter 1

INTRODUCTION

I. PREFACE

The technique of chemical ionization (CI), first introduced by Munson and Field[1] in 1966, is a direct outgrowth of fundamental studies of ion/molecule interactions; as such the technique is based on the knowledge developed from these fundamental studies and makes use of the instrumentation developed for such studies. Since its initial introduction, chemical ionization mass spectrometry (CIMS) has developed into a powerful and versatile tool for the identification and quantitation of organic molecules and, consequently, has found extensive application in many branches of chemistry and biochemistry and in medical and environmental fields.

In CIMS ionization of the sample of interest is effected by gas-phase ion/molecule reactions rather than by electron impact, photon impact, or field ionization/desorption; frequently CIMS provides information which is complementary to these techniques rather than supplementary. Particularly in structure elucidation, it is common to determine both the electron impact (EI) and chemical ionization (CI) mass spectra. Although much of the earlier work in CIMS utilized positive ion/molecule reactions, there has been, in recent years, an increased interest in negative ion/molecule reactions as an ionization method. This later development of negative ion CIMS occurred, in part, because electron impact studies at low source pressures showed, in general, a low sensitivity for negative ion production as well as a strong dependence of negative ion mass spectra on electron beam energy. In addition, many commercial mass spectrometers were not equipped to operate in the negative ion mode. This latter problem has been largely rectified, while more recent studies have shown that negative ion signals under CI conditions are at least as strong and as reproducible as positive ion signals.

The essential reactions in chemical ionization are given in general form in Reactions 1 to 3. Ionization of the reagent gas (present in large excess), frequently by electron impact, usually is followed by ion/molecule reactions involving the primary ions and the reagent gas neutrals and produces the chemical ionization reagent ion or reagent ion array (Reaction 1).

$$e + R \xrightarrow{\text{R}} R^{+/-} \tag{1}$$

$$R^{+/-} + A \longrightarrow A_1^{+/-} + N_1 \tag{2}$$

$$A_1^{+/-} \longrightarrow A_2^{+/-} + N_2 \tag{3}$$

$$\longrightarrow A_3^{+/-} + N_3$$

$$\longrightarrow A_i^{+/-} + N_i$$

This reagent ion or reagent ion array represents the stable end product(s) of the ion/molecule reactions involving the reagent gas. Collision of the reagent ion(s), $R^{+/-}$, with the additive (present at low concentration levels, usually <1% of the reagent gas) produces an ion, $A_1^{+/-}$, characteristic of the additive (Reaction 2). This additive ion may fragment by one or more pathways, as in Reaction 3, or, infrequently, react with the reagent gas; the final array of ions $A_1^{+/-}$ to $A_i^{+/-}$, when mass analyzed in the usual manner, provides the chemical ionization mass spectrum of the additive A as effected by the reagent gas R.

A large part of the usefulness of CIMS rests with the fact that a large variety of reagent gases, and hence, reagent ions, can be employed to effect ionization. To a considerable extent the choice of reagent systems can be tailored to the problem to be solved. The problems amenable to solution by the chemical ionization technique can be divided into three rough categories: (1) molecular weight determination, (2) structure elucidation, and (3) identification and quantitation.

II. MOLECULAR WEIGHT DETERMINATION

When a gaseous sample of a polyatomic species is bombarded with electrons at sample pressures of $\sim 10^{-5}$ torr, the initial electron/molecule interaction produces an assembly of molecular ions with internal energies (relative to the ground state of the molecular ion) ranging from 0 to 10 eV. For some classes of compounds the critical internal energy required for fragmentation of the molecular ion is extremely low (or zero) with the result that no molecular ions are seen in the electron impact mass spectrum; in effect fragmentation is so facile that no molecular ions survive for the $\sim 10^{-5}$ sec that elapses before mass analysis occurs. When this happens the molecular weight of the sample molecules can be established only with considerable difficulty by electron impact methods. The aim of chemical ionization in these instances is to produce, by suitable choice of reagent ions, abundant ions characteristic of the sample and providing the molecular weight of the sample molecules.

In positive ion chemical ionization the most commonly used ionization reaction for such purposes has been proton transfer (Reaction 4) from the reagent ion(s), XH^+, to the sample molecule M, where ΔH_4 is given by the proton affinity (PA) of X less the PA of M.

$$XH^+ + M \rightarrow MH^+ + X, \; \Delta H_4 \tag{4}$$

The extent of fragmentation of MH^+ depends on its internal energy which, in turn, is dependent on the exothermicity of the proton transfer reaction. The magnitude of ΔH_4, and thus the extent of fragmentation of MH^+, usually can be controlled by suitable choice of reagent gas; in addition, the stability of the even-electron MH^+ ion may be considerably greater than the stability of the odd-electron $M^{+\cdot}$ species formed in the electron impact process.

An analogous proton transfer (Reaction 5) has been utilized in negative ion chemical ionization; ΔH_5 is equal to $PA([M-H]^-) - PA(X^-)$ or, alternatively

$$X^- + M \rightarrow [M-H]^- + HX \tag{5}$$

is given by the gas-phase acidity of M less the acidity of XH; again by suitable choice of reagent ion the exothermicity of the proton transfer reaction can be made sufficiently small as to preclude extensive fragmentation of $[M-H]^-$. In addition, the exothermicity of the reaction appears to reside primarily in the bond formed between X and H and therefore is unavailable to promote fragmentation of $[M-H]^-$.

A number of other reactions also have been used to provide molecular weight information. These include hydride abstraction (Reaction 6), electron attachment (Reaction 7), electron transfer (Reaction 8), and adduct or cluster ion formation (Reaction 9).

$$X^+ + M \rightarrow [M-H]^+ + XH \tag{6}$$

$$e + M \rightarrow M^{\bar{\cdot}} \tag{7}$$

$$X^{\overline{\cdot}} + M \rightarrow M^{\overline{\cdot}} + X \qquad (8)$$

$$X^{+/-} + M \rightarrow M \cdot X^{+/-} \qquad (9)$$

It should be noted that formation of cluster ions usually requires third-body collisional stabilization and the rates of such reactions are likely to be considerably slower than the rates of bimolecular processes such as Reactions 4 to 8.

In the early literature, ions characteristic of molecular weight frequently were called quasi-molecular ions independent of their exact nature as MH^+, $[M-H]^{+/-}$, or $M \cdot X^{+/-}$. Fortunately this terminology largely seems to have been discontinued; the characteristic ions from which the molecular weight is established should be clearly identified and not hidden under a generic name.

III. STRUCTURE ELUCIDATION

When a sample of unknown chemical structure is being examined one usually desires structural information, in addition to molecular weight information, from the chemical ionization mass spectrum. Ideally, one would like an array of reagent ions which give specific characteristic reactions with each possible functional group thus signaling the presence or absence of a particular group. Except for a very few cases this ideal has not been attained, and consequently other approaches must be used.

The majority of structural studies to date have used positive ion proton transfer chemical ionization methods. The major fragmentation reactions of the even-electron MH^+ ions formed in Reaction 4 involve elimination of stable neutral molecules, HY, where Y is a functional group such as OR, NR_2, halogen, etc., present in the molecule. This fragmentation mode frequently is quite different from the fragmentation modes of the odd-electron molecular ions formed by electron impact ionization, and the structural information obtained by the two techniques often is complementary. The extent of fragmentation of MH^+ depends not only on the exothermicity of the protonation reaction but also on the nature of the functional group Y and on the identity of the fragment ion formed. In most cases, by a suitable choice of proton transfer reagent ions, both molecular weight and structural information can be obtained.

In those cases where the fragment ions formed by EI are structurally illuminating but no molecular weight information is provided by the EI mass spectrum, mixed proton transfer/charge exchange reagent systems can be used. The proton transfer reagent forms MH^+ ions indicative of molecular weight, while the charge exchange reagent, by dissociative charge exchange, gives the same fragment ions as are produced in EI since the initial ion/molecule interaction produces the odd-electron molecular ion.

The methodology of structure elucidation by negative ion chemical ionization is much less developed and impossible to generalize in an introductory chapter.

IV. IDENTIFICATION AND QUANTITATION

This application of mass spectrometry is undoubtedly the most common, whether employing electron impact or chemical ionization. The aim is to identify a known substance and to measure quantitatively the amount of this substance present in a frequently complex matrix. In CIMS three basic approaches can be used for identification. In the first, the component of interest is selectively ionized by a suitable reagent without separation, or with minimal separation, of the mixture. It is essential that the other components either not be ionized or, alternatively, that they not produce ions which are isobaric with those from the component of interest. Because of these restrictions this simple approach has been employed relatively infrequently. However, it should gain in popularity as specific ionization reactions are developed.

In the vast majority of studies separation of the mixture is achieved by gas chromatography or liquid chromatography with direct introduction of the chromatographic effluent into the mass spectrometer ion source. The ionization reaction requires only that one or more ions characteristic of the component of interest be produced in high yield to achieve maximum sensitivity of detection; these ions may be MH^+, $[M-H]^+$, $[M-H]^-$, or fragment ions. In general, formation of two or more ions is preferred since identification is made more certain by the requirements that the appropriate ion signals must show the same intensity-time profile as well as the appropriate relative intensities.

An alternative to separation of the components by chromatographic methods is separation of the components by mass spectrometric methods: so-called MS/MS experiments. In this approach the complex mixture is ionized, usually by gentle methods, to produce characteristic ions such as MH^+ or $[M-H]^-$ from each component. The ion suspected to arise from the component of interest is selected by mass analysis in the first stage of the double mass spectrometer, and its identity confirmed by mass analysis in the second stage of the instrument from the fragment ions arising from collision-induced dissociation of the selected ion.

The problems of quantitation, particularly in gas chromatography/mass spectrometry work, have been discussed in detail.[2] Because of potential losses during the workup and chromatographic steps, an internal standard usually is added to the mixture at the earliest possible stage of sample handling. This internal standard should have chemical and physical properties similar to the component of interest so that the losses of the two will be the same. Knowing the amount of standard introduced the amount of unknown present can be deduced from a comparison of the response of the mass spectrometer to the unknown and the standard. Three types of internal standards have been employed, (a) a stable isotope-labeled analogue of the compound to be measured, (b) a homologous compound which yields a fairly intense ion in common with the compound to be measured, and (c) a compound which yields a different ion but which has similar chemical and physical properties. Obviously, the type (b) standard cannot be used in the absence of temporal separation of the components by chromatography, while both type (a) and type (c) standards can be used with both chromatographic and nonchromatographic sample introduction.

V. SCOPE OF THE PRESENT WORK

These introductory comments serve to indicate the scope of the present monograph. CIMS rests firmly on the foundations established from fundamental studies of gas-phase ion chemistry. These foundations are discussed in Chapter 2. The techniques of mass spectrometry peculiar to chemical ionization studies are discussed in Chapter 3, although no attempt is made to review the basic principles of mass spectrometry; for these the reader is referred to other sources.[3-6] The reagent gases used in CIMS have been many, with various areas of utility; the more important are discussed in Chapter 4. Similarly, the range of compounds studied by chemical ionization mass spectrometry is vast. Those for which systematic studies have been made, allowing generalizations to be drawn, are discussed in Chapter 5. Finally, a number of more specialized topics are discussed in Chapter 6.

REFERENCES

1. Munson, M. S. B. and Field, F. H., Chemical ionization mass spectrometry. I. General introduction, *J. Am. Chem. Soc.,* 88, 2621, 1966.
2. Millard, B. J., *Quantitative Mass Spectrometry,* Heyden and Son, London, 1978.
3. Duckworth, H. E., *Mass Spectroscopy,* Cambridge University Press, Cambridge, 1958.
4. Kiser, R. W., *Introduction to Mass Spectrometry,* Prentice-Hall, Englewood Cliffs, N.J., 1965.
5. Roboz, J., *Introduction to Mass Spectrometry. Techniques and Applications,* Interscience, New York, 1968.
6. Melton, C. E., *Principles of Mass Spectrometry and Negative Ions,* Marcel Dekker, New York, 1970.

Chapter 2

FUNDAMENTALS OF GAS PHASE ION CHEMISTRY

I. INTRODUCTION

From the earliest days of mass spectrometry, ions were observed which generally were agreed to have arisen from reactions between ions and neutral molecules. An ion of m/z = 3 was observed by Dempster[1] in 1916 and identified as H_3^+; its formation by Reaction 1 was well established by 1925.[2,3]

$$H_2^{+\cdot} + H_2 \rightarrow H_3^+ + H^\cdot \qquad (1)$$

In 1928 Hogness and Harkness[4] reported the formation of both I_3^+ and I_3^- in iodine vapor subjected to electron impact. Ion/molecule reactions were observed in several other systems during the 1920s; the early work has been reviewed by Smyth[5] and by Thompson.[6] With improvements in instrumentation and techniques, particularly vacuum technology, the nuisance of secondary processes was largely eliminated and studies of ion/molecule reactions largely ceased. The main interests in mass spectrometry during the period 1930 to 1950 lay in the physics of ionization and dissociation, in the precise determination of isotopic masses and abundances, and in the development of analytical mass spectrometry.

The modern era of ion/molecule reaction studies began in the early 1950s when the ion CH_5^+, formed by the reaction

$$CH_4^{+\cdot} + CH_4 \rightarrow CH_5^+ + CH_3^\cdot \qquad (2)$$

was independently discovered by Tal'roze and Lyubimova[7] in the U.S.S.R. and by Stevenson and Schissler[8] and Field et al.[9] in the U.S. The observation of CH_5^+ as a stable species aroused the interest of chemists concerned with structure and bonding, while the observation that Equation 2 was considerably faster than reactions involving only neutral species suggested that ion/molecule reactions might play an important role in radiation chemistry and aroused the interest of the radiation chemists. In addition, much improved equipment was available for the controlled study of ionic collision processes. As a result a number of studies of gas-phase ion/molecule reactions were undertaken. Since that beginning the study of the products, distribution, rates, and equilibria of gas-phase ionic reactions has become a major field of scientific activity with applications in many diverse fields. In the course of these studies many advances in instrumentation have been made. The advances in instrumentation and in the understanding of gas phase ion chemistry have been reviewed in numerous articles[10-19] and books.[20-27]

The instrumentation developed for the study of ion/molecule reactions has led to the instrumentation for chemical ionization mass spectrometry; this aspect is discussed in Chapter 3. The large body of kinetic and thermochemical data derived from the fundamental studies constitutes the foundation upon which the chemistry of the chemical ionization technique is based; this body of data is reviewed in the remainder of this chapter.

II. ION/MOLECULE COLLISION RATES

The usefulness of an ion/molecule reaction in a chemical ionization system depends in part on the identity of the reaction products, and in part on the rate of the reaction.

The latter is important since only those reactions which are rapid can be expected to show adequate product ion yields. The upper limit to the reaction rate is given by the collision rate. The present section reviews the estimation from theory of collision rates of ions with nonpolar and with polar molecules.

A. Langevin Ion-Induced Dipole Theory

When an ion interacts with a nonpolar neutral molecule it induces a dipole in the neutral, the magnitude of which depends on the polarizability of the molecule; at moderately long range the only interaction of importance is the resultant ion-induced dipole interaction. Eyring et al.[28] first calculated in 1936 the classical capture collision cross section for the reaction $H_2^+ + H_2$ using absolute rate theory and a model of a structureless point charge interacting with a point-polarizable molecule. A general form of the capture collision cross section was derived by Vogt and Wannier,[29] and elaborated by Gioumousis and Stevenson[30] based on a model first developed by Langevin.[31] The theoretical development has been discussed extensively by a number of authors;[16,32-34] the present summary is based largely on the treatment of Su and Bowers.[34]

The theory calculates the collision cross section for an ion/molecule pair with a given relative velocity where both the partners are assumed to be point particles with no internal energy and the interaction between the two is assumed to arise from ion-induced dipole forces only. Thus for an ion and a molecule approaching each other with a relative velocity v and impact parameter b (Figure 1) the classical potential at an ion/molecule separation r is given by

$$V(r) = -\alpha q^2 / 2r^4 \qquad (3)$$

where α is the polarizability of the neutral and q is the charge on the ion. For $r < \infty$ the relative energy of the system, E_r, is the sum of the instantaneous kinetic energy and the potential energy

$$E_r = \frac{1}{2} \mu v^2 = E_{kin}(r) + V(r) \qquad (4)$$

where μ is the reduced mass. There are two components to $E_{kin}(r)$, the translational energy along the line of center of the collision, $E_{trans}(r)$, and the energy of relative rotation of the particles, $E_{rot}(r)$, where the latter is given by

$$E_{rot}(r) = L^2 / 2\mu r^2 = \mu v^2 b^2 / 2r^2 \qquad (5)$$

where L is the classical orbital angular momentum of the two particles. The rotational energy is associated with an outwardly acting centrifugal force and the effective potential of the ion/molecule system can be represented as the sum of the central potential energy and this centrifugal potential energy

$$V_{eff}(r) = -(\alpha q^2 / 2r^4) + (L^2 / 2\mu r^2) \qquad (6)$$

The total relative energy of the system thus is

$$E_r = E_{trans}(r) + V_{eff}(r) \qquad (7)$$

The variation of $V_{eff}(r)$ with r at constant E_r depends on the value of the impact parameter b. For b = 0 there is no contribution from the centrifugal potential and $V_{eff}(r)$ is attractive for all values of r. When b > 0 the centrifugal potential term creates a so-called "centrifugal barrier" to a capture collision. The special case (b = b_c) where the centrifugal barrier height equals E_r is shown in Figure 2; at r_c $V_{eff}(r) = E_r$ and,

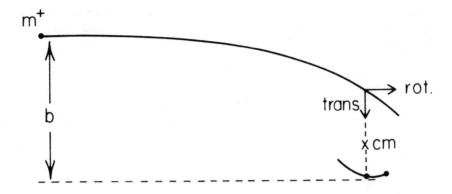

FIGURE 1. Schematic ion/molecule collision.

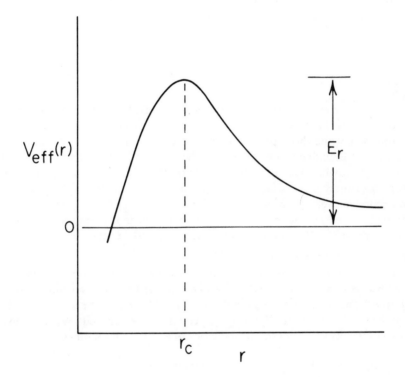

FIGURE 2. V_{eff} vs. r for critical impact parameter b_c.

thus, from Equation 7 $E_{trans}(r) = 0$ and the particles will orbit the scattering center with a constant ion/molecule separation r_c. For all impact parameters less than b_c a capture collision will occur where a capture collision is defined as one in which the particles have appropriate energy and impact parameters to pass through $r = 0$. (For real ions and molecules there would, of course, be some finite minimum value of $r > 0$.) For impact parameters greater than b_c the centrifugal barrier prevents capture and the particles are simply scattered at large values of r.

The capture collision cross section at a given velocity v is defined by

$$\sigma(v) = \pi b_c^2 \qquad (8)$$

The critical impact parameter b_c is thus the impact parameter such that $V_{eff}(r) = E_r$ and can be evaluated from the conditions that $\partial V_{eff}(r)/\partial r = 0$ and $V_{eff}(r) = E_r$ at $r = r_c$, i.e.,

$$\partial V_{eff}/\partial r = 0 = (L^2/\mu r^3) + (2\alpha q^2/r^5) \tag{9}$$

$$V_{eff}(r) = (L^2/2\mu r^2) - (\alpha q^2/2r^4) = E_r = \frac{1}{2}\mu v^2 \tag{10}$$

These restrictions lead to

$$r_c = (2q/b_c v)(\alpha/\mu)^{1/2} \tag{11}$$

$$b_c = (4q^2\alpha/\mu v^2)^{1/4} \tag{12}$$

which lead to the distance of closest approach $r_c = b_c/2^{1/2}$ and to a capture collision cross section and collision rate constant given by

$$\sigma_c(v) = \pi b_c^2 = (2\pi q/v)(\alpha/\mu)^{1/2} \tag{13}$$

$$k_c = v \cdot \sigma_c(v) = 2\pi q(\alpha/\mu)^{1/2} \tag{14}$$

Hence, if only ion-induced dipole interactions are involved, classical collision theory predicts that the capture cross section should vary inversely as the relative velocity of the colliding pair while the capture rate constant, frequently called the Langevin rate constant, should be independent of the relative velocity and the temperature. As will be seen in the following, Equation 14 predicts reasonably well the maximum rate constants for ion/molecule reactions involving nonpolar molecules. This indicates that a reaction occurs on every collision for many ion/molecule pairs; consequently there can be no activation energy for the reaction. The rate constant predicted from Equation 14 is of the order of 1×10^{-9} cm^3 molecule^{-1} sec^{-1} or 6×10^{11} ℓ mol^{-1} sec^{-1}.

B. Average Dipole Orientation (ADO) Theory

Although Equation 14 predicts reasonably well the maximum rate constants for reactions involving nonpolar molecules it underestimates the rate constants of most ion/molecule reactions involving polar molecules. For these cases it was shown first by Moran and Hamill[35] that ion-dipole forces could not be ignored. Where such forces are important the effective potential analogous to Equation 6 becomes

$$V_{eff}(r) = (L^2/2\mu r^2) - (\alpha q^2/2r^4) - (\mu_D q\cos\theta/r^2) \tag{15}$$

where μ_D is the dipole moment and θ is the angle the dipole makes with the center of collision. Hamill and colleagues[35,36] made the simplifying assumption that the dipole "locks in" on the ion ($\theta = 0$) and derived, by the approach outlined above, the capture collision rate constant

$$k_{LD}(v) = (2\pi q/\mu^{1/2})[\alpha^{1/2} + (\mu_D/v)] \tag{16}$$

where the subscript LD refers to the locked-in dipole approximation. In this case the rate constant depends on the relative velocity of the colliding pair and gives[37] for a thermal Maxwell-Boltzmann distribution of relative velocities

$$k_{LD}(therm) = (2\pi q/\mu^{1/2})[\alpha^{1/2} + \mu_D(2/\pi k_B T)^{1/2}] \tag{17}$$

where k_B is Boltzmann's constant and T is the absolute temperature.

It has been shown[37-40] that Equations 16 and 17 seriously overestimate the ion-dipole effect on ion/molecule reaction rate constants; presumably locking-in of the dipole does not occur. Thus a more realistic expression for the thermal energy rate constant is

$$k_{ADO}(\text{therm}) = (2\pi q/\mu^{1/2})\,[\alpha^{1/2} + C\mu_D(2/\pi k_B T)^{1/2}\,] \qquad (18)$$

where C, the dipole-locking constant, reflects the extent to which the dipole of the molecule orients itself with the incoming charge and may have values from 0 (no alignment) to 1 (locking-in). Bowers and Su[39.40] have developed the average dipole orientation (ADO) theory of ion/polar molecule collision in which they calculate by classical statistical theory either the average orientation angle $\bar{\theta}$ or the average $\overline{\cos}\,\theta$ (i.e., the average effective potential of the ion/polar molecule system).[41] The two approaches give similar results. In the first approach the effective potential energy at r is given by:

$$V_{eff}(r)_{ADO} = (L^2/2\mu r^2) - (\alpha q/2r^4) - (q\mu_D/r^2)\cos\bar{\theta}(r) \qquad (19)$$

which by setting $\partial V_{eff}(r)/\partial r = 0$ and $E_r = V_{eff}(r)$ at $r = r_c$ gives

$$\sigma(v) = \pi r_c^2 + (\pi q^2 \alpha/r_c^2 \mu v^2) + (2\pi q\mu_D/\mu v^2)\cos\bar{\theta}(r = r_c) \qquad (20)$$

and

$$v^2 = (q^2/\mu r_c^4) + (q\mu_D/r_c\mu)\sin\bar{\theta}(r = r_c)\left(\frac{\partial \bar{\theta}}{\partial r}\right)_{r_c} \qquad (21)$$

where r_c is the value of r at the top of the centrifugal barrier. At thermal velocities the macroscopic collision rate is given by

$$k_{ADO}(\text{therm}) = \int_0^\infty v < \sigma\,(v) > P(v)dv \qquad (22)$$

where P(v) is the normalized Maxwell-Boltzmann velocity distribution function, $\bar{\theta}(r)$ and $d\bar{\theta}/dr$ are calculated by classical statistical methods and Equation 22 evaluated by numerical integration.

From the results of such calculations Su and Bowers[39.40] deduced that the dipole-locking constant C of Equation 18 can be parameterized. It turns out that at constant temperature, C is a function of $\mu_D/\alpha^{1/2}$ only. Values of C as a function of $\mu_D/\alpha^{1/2}$ covering a temperature range of 150 to 500 K have been presented;[42] typical plots are shown in Figure 3. For large values of $\mu_D/\alpha^{1/2}$ C reaches a limiting value of ~0.26. Thus the effect of ion-dipole interactions is much less than predicted from the simple model involving locking-in of the dipole (C = 1), although enhancement of the rate constant by a factor of 2 to 4 over the ion-induced dipole value is possible for reactions where the molecule has a large dipole moment.

A number of studies,[39.40.43.44] principally involving simple proton transfer reactions, have shown that the ADO theory in the form of Equation 18 with the value of C derived from $\mu_D/\alpha^{1/2}$, adequately predicts the maximum rate constants for ion/polar molecule reactions. These will be discussed in more detail in the following sections. The ADO theory also predicts a much smaller dependence of rate constant on temperature than does the locked-dipole expression (Equation 17); this also has been confirmed experimentally.[45.46] Further extensions of theory to include conservation of an-

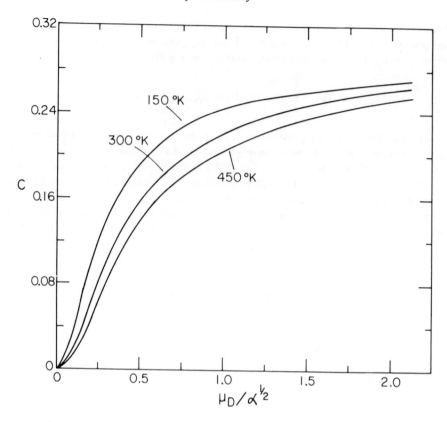

FIGURE 3.   Dipole locking constant C as a function of $\mu_D/\alpha^{1/2}$ and temperature. Data from Reference 42.

gular momentum (AADO theory)[47] and to include ion-quadrupole interactions (AQO theory)[48] have been reported. These normally are relatively small effects and the reader is referred to the original publication or to the review by Su and Bowers[34] for details.

III. POSITIVE ION/MOLECULE REACTIONS

A wide variety of ion/molecule reactions involving positive ions have been studied. The present review will concentrate primarily on the type of product ions formed and on the reaction rates. The classification system used, while essentially arbitrary, is based largely on the identity of the product ions formed.

A. Charge Exchange

The interaction of a positive ion with a neutral molecule may lead to charge exchange, Reaction 23,

$$X^{+\cdot} + M \rightarrow M^{+\cdot} + X, \Delta H = IE(M) - RE(X^{+\cdot}) \tag{23}$$

if the recombination energy (RE) of the reactant ion is greater than the ionization energy (IE) of the neutral M. The recombination energy of $X^{+\cdot}$ is defined as the exothermicity of the reaction:

$$X^{+\cdot} + e \rightarrow X, -\Delta H = RE(X^{+\cdot}) \tag{24}$$

Table 1
RECOMBINATION
ENERGIES OF
GASEOUS IONS

Ion	Recomb. en. (eV)[a]
$Ne^{+\cdot}\ ^2P_{3/2}$	21.6
$^2P_{1/2}$	21.7
$Ar^{+\cdot}\ ^2P_{3/2}$	15.8
$^2P_{1/2}$	15.9
$N_2^{+\cdot}$	15.3
$Kr^{+\cdot}\ ^2P_{3/2}$	14.0
$^2P_{1/2}$	14.7
$CO^{+\cdot}$	14.0
$CO_2^{+\cdot}$	13.8
$Xe^{+\cdot}\ ^2P_{3/2}$	12.1
$^2P_{1/2}$	13.4
$COS^{\cdot\cdot}$	11.2
$CS_2^{+\cdot}$	~9.5 or 10.0
NO^+	8.5—9.5
$C_6H_6^{+\cdot}$	9.3

[a] Data derived from References 49 and 50.

For monatomic ions the recombination energy has the same numerical value as the ionization energy of the neutral; this is not necessarily so for diatomic and polyatomic species since the product neutral in these cases may have excess vibrational/rotational energy. A selection of recombination energies taken from the summaries by Lindholm[49,50] is presented in Table 1. The exothermicity of the charge exchange reaction remains largely as internal energy of the $M^{+\cdot}$ product. If this internal energy is sufficiently large the $M^{+\cdot}$ ion may undergo fragmentation. The fragmentation reactions observed will be the same as those observed following electron impact ionization since in both cases fragmentation commences from the odd-electron molecular ion. The important difference is that the molecular ions formed by electron impact will have a distribution or range of internal energies while those formed by charge exchange will have an internal energy determined by the exothermicity of Reaction 23. Since the relative importance of the different fragmentation channels is dependent on the internal energy the product distribution from charge exchange reactions will depend on the exothermicity of the reaction (see Chapter 4, Section II.C.).

Charge exchange reactions of polyatomic molecules have been studied extensively using tandem mass spectrometers.[49,50] However, such studies have involved reactant ions with >3 eV kinetic energy and have been more concerned with product distributions than with the measurement of reaction rate constants. The information concerning the rate constants of charge exchange reactions at low kinetic energies is rather limited. Table 2 summarizes rate constants measured for charge exchange involving rare gas ions and some simple alkanes. The table also compares the measured reaction rate constants with collision rate constants calculated by Equation 14. For no case does the measured rate constant exceed the collision rate constant. With the exception of the reaction $Ne^{+\cdot} + CH_4$ the efficiency of the reactions are quite high.

Table 3 records rate constants measured for charge exchange between the ions $Ne^{+\cdot}$,

Table 2

RATE CONSTANTS FOR
CHARGE EXCHANGE
BETWEEN RARE GAS IONS
AND NONPOLAR
MOLECULES

Reactants	$k_{react}{}^a$ (Ref.)	$k_{coll}{}^a$
$Ne^{+\cdot} + CH_4$	0.07 (52)	1.26
$Kr^{+\cdot} + CH_4$	1.26 (52)	1.03
	1.03 (53)	
	1.20 (54)	
$Ar^{+\cdot} + CH_4$	1.34 (52)	1.12
	0.98 (53)	
	1.10 (54)	
$Ne^{+\cdot} + C_2H_6$	0.98 (52)	1.40
$Ar^{+\cdot} + C_2H_6$	1.07 (52)	1.16
$Kr^{+\cdot} + C_2H_6$	0.75 (52)	1.03
$Xe^{+\cdot} + C_2H_6$	0.84 (52)	0.98
$Ne^{+\cdot} + C_3H_8$	1.21 (52)	1.55
$Ar^{+\cdot} + C_3H_8$	1.20 (52)	1.25

^{a} Rate constants: cm^3 molecule^{-1} sec^{-1} × 10^9.

Table 3

RATE CONSTANTS FOR CHARGE EXCHANGE
REACTIONS INVOLVING POLAR MOLECULES

Reactants	$k_{calc'd}{}^a$			$k_{expt'l}{}^{a,b}$
	Ion-ind. dipole	Ion-dipole	Total	
$Ne^{+\cdot} + (CH_3)_2O$	1.44	0.51	1.95	1.66
$Ne^{+\cdot} + CH_3OH$	1.21	0.93	2.14	1.36
$Ne^{+\cdot} + (CH_3)_2CO$	1.54	1.52	3.06	3.15
$Ar^{+\cdot} + (CH_3)_2O$	1.16	0.42	1.58	1.48
$Ar^{+\cdot} + CH_3OH$	1.01	0.77	1.78	1.55
$Ar^{+\cdot} + (CH_3)_2CO$	1.22	1.21	2.43	2.56
$N_2^{+\cdot} + (CH_3)_2O$	1.29	0.46	1.75	1.95
$N_2^{+\cdot} + CH_3OH$	1.10	0.84	1.94	1.75
$N_2^{+\cdot} + (CH_3)_2CO$	1.37	1.35	2.72	2.76
$Kr^{+\cdot} + (CH_3)_2O$	0.99	0.35	1.34	1.43
$Kr^{+\cdot} + CH_3OH$	0.88	0.68	1.56	1.72
$Kr^{+\cdot} + (CH_3)_2CO$	1.02	1.00	2.02	1.79
$CO^{+\cdot} + (CH_3)_2O$	1.29	0.46	1.75	1.94
$CO^{+\cdot} + CH_3OH$	1.10	0.84	1.94	1.83
$CO^{+\cdot} + (CH_3)_2CO$	1.37	1.35	2.72	2.83

^{a} Rate constants: cm^3 molecule^{-1} sec^{-1} × 10^9.
^{b} Uncertainty ±0.2, data from Reference 55.

$Ar^{+\cdot}$, $N_2^{+\cdot}$, $Kr^{+\cdot}$, and $CO^{+\cdot}$ and the polar molecules dimethyl ether (μ_D = 1.30D), methyl alcohol (μ_D = 1.71D), and acetone (μ_D = 2.88D). Also included in the table are the collision rate constants calculated from the ADO theory in Equation 18. The latter have been separated into the ion-induced dipole and ion-permanent dipole contributions to illustrate the magnitude of each. The ion-permanent dipole contribution varies with the dipole moment of the neutral, and with the most polar molecule, acetone, is equal in magnitude to the ion-induced dipole contribution. Again, within experimental error, the measured reaction rate constants do not exceed the collision rate constants although in most cases the reaction efficiency is high.

In principle the electron transfer involved in Reaction 23 could occur over larger distances than are implied in capture collisions; as a consequence the observed reaction rate constants could be larger than the calculated capture collision rate constants. Early rate constant measurements for charge exchange involving methane[52,56] and silane[52] neutrals which supported this idea have not been reproduced in later work.[53] Bowers and co-workers[38,39] have reported reaction rate constants for charge exchange involving rare gas ions and dichloroethylenes and difluorobenzenes which are considerably larger than the calculated capture collision rate constants; they have interpreted their results in terms of a long-range electron jump. The results in Tables 2 and 3 provide no support for the long-range electron jump mechanism and with the limited data available it is not clear how prevalent this mechanism will be in charge exchange reactions. It does appear from the data available that the rate constants for most charge exchange reactions will be close to the capture collision limit.

B. Proton and Hydrogen Atom Transfer

The self-protonation reaction

$$M^{+\cdot} + M \rightarrow MH^+ + [M - H]^{\cdot} \tag{25}$$

is an ubiquitous reaction for many classes of compounds. Indeed, it is so prevalent that in many cases MH^+ ions may be observed under electron impact conditions unless precautions are taken. (Fragment ions produced by electron impact also may react to form MH^+.) Reaction 25 may occur either by transfer of H^+ from $M^{+\cdot}$ to M or by transfer of an H atom from M to $M^{+\cdot}$; the two reactions cannot be distinguished in a conventional single-source experiment. Using a tandem mass spectrometer Abramson and Futrell[61] studied the reaction between $CD_4^{+\cdot}$ and CH_4. They found that D^+ transfer was favored over H atom transfer by approximately a factor of 2. Similar results showing that H^+ transfer is favored over H atom transfer have been obtained by Huntress and Pinnizotto[57] for the reaction pairs $H_2O^{+\cdot} + H_2O$ and $NH_3^{+\cdot} + NH_3$ using isotopic labeling and ICR techniques. A further interesting observation from the $CD_4^{+\cdot} + CH_4$ experiment is that only CD_4H^+ and CH_4D^+ product ions are observed with no isotopically mixed species such as $CH_3D_2^+$. This result indicates that if a collision complex is formed it is not sufficiently long-lived for isotopic interchange between reaction partners to occur. This appears to be generally true for H^+/H-atom transfer reactions and contrasts with the results obtained for condensation reactions which are discussed in Section III.D.

Table 4 records the measured reaction rate constants for a number of self-protonation reactions and compares these with the calculated capture collision rate constants. With the exception of the first entry, these were calculated from the ADO theory Equation 18. The rate constant for the reaction between $NH_3^{+\cdot}$ and NH_3 is known to depend on the vibrational state of the reactant ion.[57] In general, reaction occurs on 50 to 90% of collisions for this type of reaction. These reactions are important in the preparation of gaseous Brønsted acid reagent ions.

Of even more interest in the present context are proton transfer reactions from gaseous Brønsted acids to neutral molecules, as in Reaction 26; this is a common ionization reaction in chemical ionization mass spectrometry.

$$BH^+ + M \rightarrow MH^+ + B \tag{26}$$

Table 4
REACTION RATE CONSTANTS FOR SELF-PROTONATION REACTIONS $M^{+\cdot} + M \rightarrow MH^+ + [M-H]^\cdot$

$M^{+\cdot} + M$	k_{react}^a (Ref.)	k_{coll}^a
$CH_4^{+\cdot} + CH_4$	1.11 (57)	1.35
$NH_3^{+\cdot} + NH_3$	1.5—2.2 (57)	2.10
$H_2O^{+\cdot} + H_2O$	2.05 (57)	2.29
$CH_3OH^{+\cdot} + CH_3OH$	2.53 (37)	1.78
$CH_3NH_2^{+\cdot} + CH_3NH_2$	1.22 (58)	1.85
$CH_3F^{+\cdot} + CH_3F$	1.87 (59)	1.89
$CH_3Cl^{+\cdot} + CH_3Cl$	1.53 (59)	1.63
$CH_3SH^{+\cdot} + CH_3SH$	0.77 (60)	1.43

^{a} Rate constant: cm^3 molecule^{-1} sec$^{-1} \times 10^9$.

Table 5
H^+/D^+ TRANSFER FROM GASEOUS BRØNSTED ACIDS

Reactants	k_{react} (Ref.)^a	k_{coll}^a
$CD_5^+ + C_2H_6$	1.07 (62)	1.39
$CH_5^+ + C_3H_8$	1.54 (63)	1.68
$CH_5^+ + C_2H_4$	1.51 (64)	1.49
$CH_5^+ + CH_3NH_2$	2.51 (44)	1.98
	2.25 (40)	
$CH_5^+ + (CH_3)_2NH$	2.25 (44)	1.93
	2.15 (40)	
$CD_5^+ + CH_3Cl$	2.95 (44)	2.27
$CH_5^+ + C_2H_5Cl$	3.02 (65)	2.66
$CH_5^+ + n\text{-}C_5H_{11}Cl$	3.29 (65)	3.03
$CD_5^+ + CH_3OH$	1.66 (44)	1.98
$C_2H_5^+ + CH_3NH_2$	1.82 (44)	1.70
	1.87 (40)	
$C_2H_5^+ + (CH_3)_2NH$	1.83 (44)	1.62
	1.88 (40)	
$C_2D_5^+ + CH_3SH$	1.96 (60)	1.74
$C_3H_7^+ + CH_3NH_2$	1.65 (40)	1.63
$C_3H_7^+ + (CH_3)_2NH$	1.64 (40)	1.40
$C_4H_9^+ + CH_3NH_2$	1.43 (66)	1.54
$C_4H_9^+ + (CH_3)_2NH$	1.38 (66)	1.40

^{a} cm^3 molecule^{-1} sec$^{-1} \times 10^9$.

Table 5 records experimentally measured rate constants for proton transfer from CH_5^+, $C_2H_5^+$, $C_3H_7^+$, and $C_4H_9^+$ to a variety of small molecules and compares these rate constants with the calculated capture collision rate constants. The first three entries represent proton transfer to nonpolar molecules with the collision rate constants calculated from the polarization theory by Equation 14; the remaining entries represent proton transfer to polar molecules with the collision rate constants calculated from the ADO theory using appropriate values of C in Equation 18.

These exothermic proton transfer reactions are all highly efficient with the reaction occurring on essentially every collision. Indeed, the ADO theory appears to slightly ($\sim$20%) underestimate the collision rate. Better agreement is obtained using the aver-

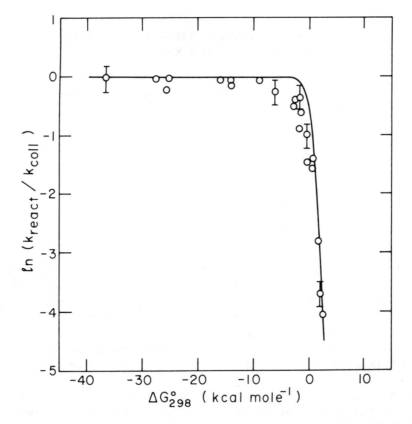

FIGURE 4.   Efficiency of proton transfer as a function of ΔG°_{react}.   (From Bohme, D. K., Mackay, G. I., and Schiff, H. I., *J. Chem. Phys.*, 73, 4976, 1980. With permission.)

age dipole orientation theory with conservation of angular momentum (AADO theory); this is discussed by Su and Bowers.[34] The high efficiency of exothermic proton transfer reactions also has been established in other studies using a variety of Brønsted acids.[43,67,68] Although all of these studies have involved rather simple neutral reactants there is no reason to believe that proton transfer to complex molecules will not be equally efficient. For such complex molecules with high polarizabilities and, frequently, dipole moments proton transfer rate constants in the range 2 to 4×10^{-9} cm^3 molecule^{-1} sec^{-1} can be expected.

As the proton transfer reaction becomes thermoneutral the reaction efficiency drops[69,70] and becomes very low for endothermic or endoergic reactions. This is illustrated by the plot in Figure 4 of the reaction efficiency (k_{react}/k_{coll}) as a function of ΔG°_{react} for a variety of proton transfer reactions as reported by Bohme et al.[70] The moral is that for maximum ionization efficiency in proton transfer CI one requires an exothermic protonation reaction. However, the more exothermic the protonation reaction the more extensive will be the fragmentation of MH$^+$. This aspect is discussed in detail in Chapter 4.

C. Negative Ion Transfer Reactions

The hydride ion transfer reaction

$$X^+ + M \rightarrow [M - H]^+ + XH \tag{27}$$

Table 6

HYDRIDE ION ABSTRACTION REACTIONS OF ALKANES $X^+ + M \rightarrow XH + [M-H]^+$

$k(cm^3 \ molecule^{-1} \ sec^{-1}) \times 10^{10}$

M /X =	$C_2H_5^+$	$i\text{-}C_3H_7^+$	$t\text{-}C_4H_9^+$	CF_3^+	NO^+
C_3H_8	6.3	1.2	—	5.9	—
$n\text{-}C_4H_{10}$	8.4	5.6	—	7.5	<0.02
$i\text{-}C_4H_{10}$	10	4.2	—	5.5	4.6
$n\text{-}C_5H_{12}$	10.9	8.3	<0.004	8.9	<0.05
$i\text{-}C_5H_{12}$	10.9	7.5	0.25	8.0	7.8
$c\text{-}C_6H_{12}$	16	11	<0.0005	—	1.3
$2\text{-}CH_3C_5H_{11}$	—	—	0.39	—	11.8
$3\text{-}CH_3C_5H_{11}$	—	—	0.27	—	12.1
$n\text{-}C_7H_{16}$	—	—	—	11.8	0.47
$3\text{-}CH_3C_6H_{13}$	—	—	0.45	—	14.8

was first discovered by Field and Lampe[71] in a study of the ion/molecule reactions occurring on electron impact ionization of the C_2 to C_6 alkanes; in these systems the reactant ions were the lower alkyl ions produced by dissociative ionization of the alkane. Such reactions lead to the formation of the $C_4H_9^+$ reactant ion as the major species in the high pressure mass spectrum of i-butane (see Chapter 4, Section II.A.). They also are of importance in the radiolysis of hydrocarbons and have been studied extensively in radiolytic and mass spectrometric experiments by Ausloos and co-workers.[25,72,73]

Table 6 presents a summary of rate constants for hydride abstraction from alkanes by various reactant ions. The majority of these results come from mass spectrometric experiments[74,75] although some of the results for $t\text{-}C_4H_9^+$ come from radiolysis experiments.[25,76,77] In a few cases there are differences of up to a factor of 2 in reported rate constants for the same reaction. In all cases hydride ion transfer is the only reaction observed and calculated collision rate constants are in the range 13 to 16 × 10^{-10} cm^3 molecule^{-1} sec^{-1}. The $C_2H_5^+$ and $s\text{-}C_3H_7^+$ ions react with the higher n-alkanes and branched alkanes with rate constants approaching the calculated collision rate constants. By contrast the $t\text{-}C_4H_9^+$ ion is essentially unreactive with n-alkanes and shows a low reactivity (<6% of collision rate) with branched alkanes. The NO^+ ion shows a low reactivity with n-alkanes but a high reactivity, approaching the collision rate, with branched alkanes. The CF_3^+ ion is quite reactive with both normal and branched alkanes showing reaction rate constants approaching the collision rate constants for the higher alkanes.

Thermochemical effects appear to be playing a significant role in these reactions. Table 7 summarizes the thermochemistry for H^- abstraction from specific (primary, secondary, tertiary) positions of n-butane and i-butane by the reactant ions of Table 6. Also included in the table are the % H^- abstraction from the specific positions as established by deuterium labeling. Reactions of $C_2H_5^+$ or CF_3^+ to abstract H^- from primary, secondary, or tertiary positions are all exothermic and reaction of all sites is observed. On the other hand reaction of $s\text{-}C_3H_7^+$ to abstract H^- from a primary position is endothermic and only abstraction from secondary or tertiary positions is observed. Within reasonable error limits only abstraction of H^- from tertiary positions is thermochemically permitted with NO^+ as a reactant; consequently NO^+ reacts at a significant rate only with those alkanes possessing a tertiary hydrogen (in contrast to these results the NO CI of n-alkanes is reported[78] to produce $[M-H]^+$ although the sensitivity of the method has not been noted). The reaction of $t\text{-}C_4H_9^+$ to abstract the tertiary hydrogen from i-butane is thermoneutral while reaction with more complex branched

Table 7

THERMOCHEMISTRY OF HYDRIDE ABSTRACTION FROM
n-C$_4$H$_{10}$ AND t-C$_4$H$_{10}$[a]

| | ΔH_{react} (% of total abstraction) | | | |
	C$_2$H$_5^+$	C$_3$H$_7^+$	CFCF$_3^+$	NO$^+$
n-C$_4$H$_{10}$(prim)	− 8 (—)	+ 14 (0)	−35 (36)	+ 18 (—)
n-C$_4$H$_{10}$ (sec)	−26 (—)	− 4 (100)	−53 (64)	0 (—)
i-C$_4$H$_{10}$(prim)	− 8 (68)	+ 14 (0)	−35 (76)	+ 8 (0)
i-C$_4$H$_{10}$(tert)	−41 (32)	− 19 (100)	−68 (24)	− 15 (100)

[a] Thermochemical data from Table 21, % abstraction data from Reference 75.

alkanes is only slightly exothermic. Consequently it is not surprising that t-C$_4$H$_9^+$ reacts relatively slowly; detailed consideration[75] of the results also suggests that steric effects may play a role in the reactions of t-C$_4$H$_9^+$, while for the remaining systems there is some evidence for a dependence of the reaction rate on the lifetime of the ion/neutral collision complex.

Although quantitative data are restricted to the alkanes, hydride abstraction from other classes of molecules is known. For example, in chemical ionization studies Hunt and Ryan[79] have observed hydride ion abstraction from alcohols and aldehydes by NO$^+$. In fact, hydride ion abstraction is a possible reaction whenever Reaction 27 is exothermic, i.e., whenever the hydride ion affinity (HIA) of X$^+$ is greater than the HIA of (M-H)$^+$. Hydride ion affinities are discussed below in Section VI.B. and a tabulation of hydride ion affinities is given in Table 21.

Olefinic ions also abstract H$^-$ from alkanes, but usually also react by H$_2^-$ abstraction, thus converting the olefinic ion to a neutral saturated hydrocarbon.[80-84] Thus, for example, the deuterated propene molecular ion reacts with isopentane by both Reactions 28 and 29

$$CD_3CDCD_2^{+\cdot} + i - C_5H_{12} \rightarrow CD_3CDCD_2H + C_5H_{11}^+ \qquad (28)$$

$$\rightarrow CD_3CDHCD_2H + C_5H_{10}^{+\cdot} \qquad (29)$$

with the ratio $k_{28}/k_{29} = 0.55$. The H$_2^-$ transfer reaction has found no practical application in chemical ionization mass spectrometry. Halide ion transfer reactions (Reaction 30, Y = halogen)

$$R_1^+ + R_2Y \rightarrow R_2^+ + R_1Y \qquad (30)$$

also have been observed[85-88] and may be of importance in some chemical ionization systems although the rates tend to be low.

D. Condensation Reactions

In a condensation ion/molecule reaction a relatively long-lived collision complex is formed between the two reactants. This complex, under suitable conditions, may be collisionally stabilized (Reaction 31) but in the absence of such stabilization decomposes, at least in part, to products different from the reactants (Reaction 32). This type of reaction is particularly

$$A^+ + B \rightarrow [AB^+]^* \xrightarrow{M} AB^+ \qquad (31)$$

$$\longrightarrow C^+ + D \qquad (32)$$

Table 8
ION/MOLECULE REACTIONS IN C_4H_8
OLEFINS[a] $C_4H_8^{+\cdot}$ + C_4H_8 → PRODUCTS

	Yield (% of total reaction)			
Product	i-C_4H_8	1-C_4H_8	2-C_4H_8	c-$C_3H_5CH_3$
$C_4H_9^+$	91.7	8.3	3.3	<1.0
$C_5H_9^+$	—	4.3	—	8.8
$C_5H_{10}^+$	4.5	38.8	5.8	69.0
$C_5H_{11}^+$	—	4.2	3.5	—
$C_6H_{11}^+$	3.4	26.3	42.3	8.0
$C_6H_{12}^+$	—	16.1	45.1	7.6
k_{react}[b]	5.4	6.0	0.37	0.60

[a]   From Reference 90.
[b] $\times 10^{10}$ cm^3 molecule^{-1} sec^{-1}.

prevalent in the gas-phase ion chemistry of alkenes and alkynes, involving reaction of both the molecular ion and fragment ions with the neutral, and normally leads to products of higher carbon content than the reactant ion. The total rate of reaction and the distribution of product ions formed depend strongly on the identity of the reactants. As an example, Table 8 summarizes the product distribution from reaction of the $C_4H_8^{+\cdot}$ molecular ions (formed from the respective neutral olefins) with the C_4H_8 olefins listed; also included in the table are the total rate constants for reaction. These are, in general, lower than the collision rate constant of ~1.3 × 10^{-9} cm^3 molecule^{-1} sec^{-1}. In the isobutene system the major part of the reaction occurs by H$^+$/H-atom transfer to yield the $C_4H_9^+$ product, but with the remaining systems a variety of C_5 and C_6 products are formed. In many systems at higher pressure the collision complex is stabilized by collision and is observed.

The H$^+$/H-atom transfer reaction appears to occur by a different mechanism than the condensation reactions leading to the products of higher carbon number. Deuterium-labeling results[90-93] show that the former reaction occurs by a "direct" mechanism in which there is no isotopic interchange between the reacting partners while the condensation reactions occur through formation of a long-lived intimate complex in which isotopic interchange between the reaction partners does occur. As an example, in the reaction of $C_3H_6^{+\cdot}$ (propylene) with CD$_2$ = CDCD$_3$ only $C_3H_6D^+$ and $C_3D_6H^+$ are observed as $C_3(H,D)_7^+$ products while all possible $C_5(H,D)_9^+$ ions from $C_5H_6D_3^+$ to $C_5H_3D_6^+$ are observed.[90] Evidence has been presented[89,91,94-99] that an increase in the energy (internal or kinetic) of the reactants increases the importance of the H$^+$/H-atom transfer reaction at the expense of the condensation reactions.

Considerable interest has centered on the structures of the long-lived complexes formed in condensation ion/molecule reactions.[25] Of particular relevance in the present context is the evidence for four-centered cyclic complexes in the reactions of propylene,[92] allene and propyne,[100] and the fluoroethylenes.[101] Similar four-centered complexes have been invoked to rationalize the results obtained using vinyl methyl ether as a CI reagent gas for the determination of double bond position in olefinic compounds.[102,103]

Condensation ion/molecule reactions play only a minor role in chemical ionization studies. The most familiar example is Reaction 33

$$CH_3^+ + CH_4 \rightarrow C_2H_5^+ + H_2 \tag{33}$$

forming the $C_2H_5^+$ reactant ion in methane at high pressure. At low ion kinetic energies

Table 9
ASSOCIATION REACTIONS OF POSITIVE IONS

Reaction	k (cm^6 molecule^{-2} sec^{-1})	Ref.
$Ar^+ + 2Ar \rightarrow Ar_2^+ + Ar$	3×10^{-31}	105
$N_2^+ + 2N_2 \rightarrow N_4^+ + N_2$	8×10^{-29}	106
$CO_2^+ + 2CO_2 \rightarrow (CO_2)_2^+ + CO_2$	3×10^{-28}	107
$H_3O^+ + H_2O + N_2 \rightarrow H^+(H_2O)_2 + N_2$	3.4×10^{-27}	106
$H^+(H_2O)_2 + H_2O + N_2 \rightarrow H^+(H_2O)_3 + N_2$	2.3×10^{-27}	106
$H^+(H_2O)_3 + H_2O + N_2 \rightarrow H^+(H_2O)_4 + N_2$	2.4×10^{-27}	106
$NH_4^+ + NH_3 + O_2 \rightarrow H^+(NH_3)_2 + O_2$	1.8×10^{-27}	108

beam experiments studying the $CD_3^+ + CH_4$ reaction have shown that a long-lived complex is formed in which interchange of H and D does occur;[52] however the complex is not sufficiently long-lived for $C_2H_7^+$ to be collisionally stabilized. The reaction rate constant for Reaction 33 $\sim 1.15 \times 10^{-9}$ cm^3 molecule^{-1} sec^{-1} essentially equal to the calculated collision rate constant of 1.22×10^{-9} cm^3 molecule^{-1} sec^{-1}.[104] The collisionally stabilized complexes of ion/molecule reactions frequently are observed at the pressures used in CI experiments. Thus the $[M + C_2H_5]^+$ and $[M + C_3H_5]^+$ ions observed in CH_4 CI mass spectra can be considered as examples of stabilized complexes of condensation ion/molecule reactions. In the absence of stabilization fragmentation of the complex may occur; these fragmentation reactions are discussed in Chapter 4.

E. Clustering or Association Reactions

$$A^+ + B \xrightarrow{M} AB^+ \tag{34}$$

In the association or clustering Reaction 34 the collision of an ion with a molecule results in formation of a complex which is observed after collisional stabilization. However, in contrast to the condensation reactions discussed above, in the absence of collisional stabilization the complex dissociates to re-form the reactants rather than to form new products. A particularly common type of clustering reaction is the solvation of gas-phase ions by polar molecules.

As the data in Table 9 show, the third-order rate constants for clustering reactions cover the range from $\sim 10^{-31}$ to $> 10^{-27}$ cm^6 molecule^{-2} sec^{-1}; in general the rate constant increases with an increase in the number of internal degrees of freedom of the collision complex. The higher values correspond to an effective second-order rate constant at 1 torr of $\sim 10^{-11}$ cm^3 molecule^{-1} sec^{-1}. While this is considerably smaller than many of the rate constants quoted in earlier sections, clustering reactions still can play a significant role in CI studies when polar species capable of hydrogen bonding are employed. As an illustration Figure 5 shows the ions observed in methylamine, ionized by 10.0 eV photons, as a function of ion source pressure.[58] At pressures >0.1 torr a variety of proton-bound cluster ions are observed. This behavior generally will be true for compounds, such as alcohols or amines, capable of forming proton-bound cluster ions. The thermochemical stability of such solvated ions increases with the extent of clustering,[18,105] consequently the proton affinity increases with increasing solvation and the exothermicity of proton transfer decreases. Hence, the chemical ionization mass spectra may be much more strongly dependent on reagent gas source pressure than normally is the case for nonclustering reagent species. The extent of formation of cluster ions can be reduced by diluting the polar reagent gas with a nonpolar gas; however, even under these conditions solvation of the product ions may occur. In addition, for polar sample molecules, formation of species such as M_2H^+ may occur unless the sample pressure is kept quite low.

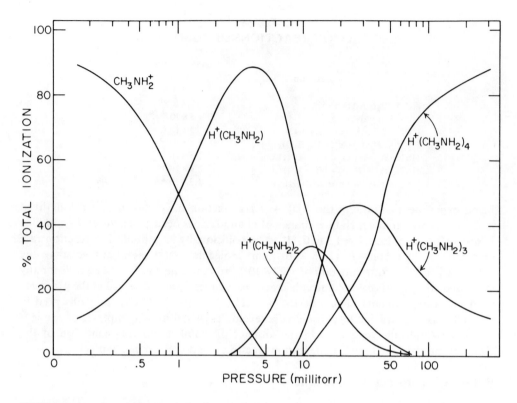

FIGURE 5.   Ion intensities in methyl amine as a function of source pressure. (From Hellner, L. and Sieck, L. W., *Int. J. Chem. Kin.*, 5, 177, 1973. With permission.)

IV. NEGATIVE ION/MOLECULE REACTIONS

Reactions between negative ions and neutral molecules have been studied much less extensively and systematically than positive ion/molecule reactions. In addition, many of the detailed studies which have been made have involved simple species of interest with regard to the ion chemistry of the Earth's atmosphere;[109] such species normally are not of major interest in chemical ionization studies. The following discussion attempts to review those aspects of negative ion/molecule chemistry which are of relevance to chemical ionization. Although, in principle, electron/molecule interactions fall outside the scope of such a review, electron attachment is an important mode of formation of negative ions which frequently is used in chemical ionization mass spectrometry; therefore, electron/molecule interactions will be reviewed.

A. Electron/Molecule Interactions[110,111]

Negative ions are produced as a result of electron/molecule interactions by three general processes depicted for the molecule MX

1.   Ion-pair formation:

$$e + MX \rightarrow M^+ + X^- + e \tag{35}$$

2.   Electron attachment:

$$e + MX \rightarrow MX^- \tag{36}$$

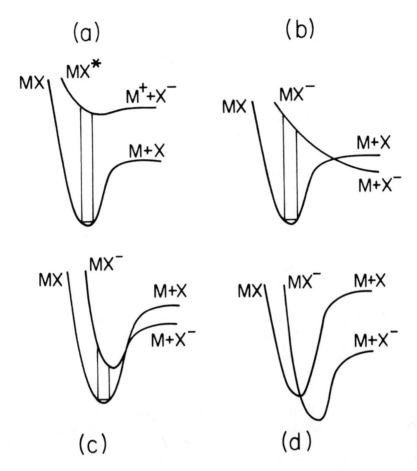

FIGURE 6. Potential energy curves for negative ion formation.

3. Dissociative electron attachment:

$$e + MX \rightarrow M + X^- \tag{37}$$

These processes may be better understood with reference to potential energy diagrams depicted in Figure 6 for a hypothetical diatomic molecule; similar arguments will apply to polyatomic molecules if potential energy surfaces are considered. In the following discussion the electron affinity is defined as the energy of the ground state neutral and an electron at infinite distance minus the energy of the ground state of the negative ion.

In ion-pair formation, Figure 6a, the electron provides the energy necessary to form an excited molecule which dissociates (or predissociates) to give a positive ion and a negative ion. The thresholds for such processes usually lie above 10 to 15 eV electron energy and the cross section increases approximately linearly with excess electron energy to an energy roughly three times the threshold energy. Negative ion formation by ion-pair formation does not appear to be a particularly common process.

Dissociative electron attachment occurs (Figure 6b) if capture of an electron leads by a vertical Franck-Condon transition to a repulsive state of $Mx^{\mp}$ which dissociates to form $M^{\cdot} + X^-$, with possible excess internal and translational energy.

For electron attachment two cases arise depending on whether the electron affinity of MX is <0 (Figure 6c) or >0 (Figure 6d). When the electron affinity is less than 0 a

Table 10
LIFETIMES OF
MOLECULAR ANIONS[a]

M^-	t (μsec)
$C_6H_5NO_2^-$	$\sim$40
Phthalic anhydride$^-$	313
o-$ClC_6H_4NO_2^-$	17
o-$C_6H_4(NO_2)_2^-$	463
m-$NO_2C_6H_4COCH_3^-$	310
o-$CH_3OC_6H_4NO_2^-$	16
$C_6H_5F^-$	$<1 \times 10^{-6}$
$C_6F_6^-$	12
$C_6H_6^-$	1×10^{-6}
$C_6H_5CN^-$	5
$C_6F_5CN^-$	17
O_2^-	2×10^{-6}

[a]   From Reference 112.

Franck-Condon transition leads to an unstable molecular anion [MX$^-$]* which may disappear by autodetachment (Reaction 38) or if above the dissociation limit, may dissociate to M$^{\cdot}$ + X$^-$.

$$[MX^-]^* \rightarrow MX + e \qquad (38)$$

Autodetachment in this situation usually is very rapid and occurs within a vibrational period leaving little possibility for collisional stabilization of [MX$^-$]* unless the pressure is very high. Stabilization by radiation emission also is possible but not probable.

When the electron affinity of MX is greater than 0 (Figure 6d) the potential energy curve for MX$^-$ may or may not intersect the vertical Franck-Condon region. If it does the situation is similar to that discussed with respect to Figure 6c and similar reactions occur. If it does not a different form of attachment is imagined. Essentially, because of the low energy (thermal) of the electron, long interaction times result. Under these conditions the Born-Oppenheimer approximation may break down and the nuclei relax and change to a position on the negative ion curve. If the excess energy of [MX$^-$]* can be distributed into the vibrational modes of the anion, significant lifetimes (with respect to autodetachment) for the molecular anion may be observed and collisional stabilization can occur more readily. All of the electron attachment processes are resonance processes since no electron is produced to carry away the excess energy. They are observed from thermal up to <15 eV electron energy, with thermal energies usually corresponding to electron attachment and higher energies corresponding to dissociative electron attachment.

The molecular anions formed by electron attachment are inherently unstable with respect to autodetachment (Reaction 38) and are stabilized by radiation emission or by collisional deactivation. For significant collisional stabilization at pressures of $\sim$1 torr (CI pressures) the natural lifetime of the [MX$^-$]* species formed by electron attachment must be >1 μsec. Table 10 summarizes lifetimes measured[112] for a number of anions formed by electron attachment; such lifetimes range from <10^{-12} sec to >10^{-4} sec. Ions with long lifetimes are molecules with positive electron affinities (often conjugated systems bearing electron-attracting substituents) and generally are large molecules where the excess energy of [MX$^-$]* can be distributed among the many internal degrees of freedom to delay autodetachment. In a number of cases the lifetimes are

Table 11
ELECTRON ATTACHMENT
RATE CONSTANTS

M	k (cm^3 molecule^{-1} sec^{-1})
CH_3Cl	$\sim 5.6 \times 10^{-11}$
CH_2Cl_2	1.4×10^{-11}
$CHCl_3$	3.6×10^{-9}
CCl_4	2.5×10^{-7}
CH_3CCl_3	1.5×10^{-8}
$CH_2ClCHCl_2$	1.4×10^{-10}
CF_3Cl	4.3×10^{-10}
CF_3Br	1.2×10^{-8}
CF_2Br_2	2.4×10^{-7}
CF_4	7.1×10^{-13}
$C_6H_5NO_2$	5.8×10^{-10}
C_6F_6	9.0×10^{-8}
$C_6F_5CF_3$	2.2×10^{-7}
Azulene ($C_{10}H_8$)	2.6×10^{-8}

significantly greater than 10 μsec and molecular anions should be observable in low-pressure mass spectrometry where collisions are negligible; this has been found to be so for a variety of compounds using derivatives containing conjugated dicarbonyl units[113,114] or nitrophenyl groups.[115,116]

A wide range of electron attachment rate coefficients also has been reported. Table 11 summarizes data for a variety of simple halogenated alkanes[117] and some aromatic molecules.[112] For the halogenated alkanes the negative ions observed are halide anions while for the aromatic molecules the molecular anions are observed. Of particular note is the fact that a number of the attachment rate coefficients are in the range 10^{-8} to 10^{-7} cm^3 molecule^{-1} sec^{-1}, considerably higher than the maximum rate coefficients for ion/molecule reactions; these high rate coefficients are a result of the high mobility of the electron. As a result of these high attachment rate coefficients, in favorable circumstances electron capture chemical ionization can have a considerably higher sensitivity than chemical ionization techniques relying on ion/molecule reactions. This will occur if the sample molecule is derivatized so that high attachment rate coefficients and long-lived molecular anions result.[118,119]

B. Associative Detachment Reactions

Negative ions may disappear from the system by the associative detachment reaction

$$X^- + N \rightarrow XN + e \qquad (39)$$

where N is a neutral species. Such reactions have been studied primarily in simple systems of interest in atmospheric chemistry.[120] Table 12 summarizes some rate constants reported[120-122] for a number of reactions involving the O^- ion chiefly. In a number of cases reaction rate constants approaching collision rate constants are observed. These reactions lead to a loss of negative ion signal. The prevalence of associative detachment reactions in the more complex systems applicable to chemical ionization has not been studied; however, it is a complication which should be borne in mind since it can lead to a decrease in sensitivity.

C. Displacement Reactions

The displacement reaction

$$\dot{X}^- + RY \rightarrow RX + Y^- \qquad (40)$$

Table 12
RATE CONSTANTS FOR
ASSOCIATIVE DETACHMENT
REACTIONS

Reaction	$k \times 10^{10}$ [a]	Ref.
$O^{\overline{\cdot}} + C_2H_2 \rightarrow C_2H_2O + e$	13	120
$O^{\overline{\cdot}} + C_2H_4 \rightarrow C_2H_4O + e$	4.1	120
$O^{\overline{\cdot}} + H_2 \rightarrow H_2O + e$	7.0	120
$O^{\overline{\cdot}} + CO \rightarrow CO_2 + e$	6.5	120
$H^- + O_2 \rightarrow HO_2^{\overline{\cdot}} + e$	12.0	120
$O^{\overline{\cdot}} + NO \rightarrow NO_2 + e$	5	121
$OH^- + H \rightarrow H_2O + e$	10	122

[a] cm^3 molecule^{-1} sec^{-1}.

Table 13
RATE COEFFICIENTS FOR DISPLACEMENT
REACTIONS $X^- + RY \rightarrow RX + Y^-$

X^-	RY	$k \times 10^9$ cm^3 molecule^{-1} sec^{-1}	React. efficiency
OH^-	CH_3Cl	1.6	0.68
F^-	CH_3Cl	0.8	0.35
CH_3O^-	CH_3Cl	0.49	0.25
CH_3S^-	CH_3Cl	0.078	0.045
CN^-	CH_3Cl	<0.001	<0.0005
OH^-	CH_3Br	1.9	0.84
F^-	CH_3Br	0.6	0.28
CH_3O^-	CH_3Br	0.72	0.40
CH_3S^-	CH_3Br	0.14	0.091
Cl^-	CH_3Br	0.012	0.007
CN^-	CH_3Br	0.02	0.01
OH^-	$CF_3CO_2CH_3$	1.4	0.47
F^-	$CF_3CO_2CH_3$	1.1	0.39
CH_3O^-	$CF_3CO_2CH_3$	1.0	0.43
CH_3S^-	$CF_3CO_2CH_3$	0.50	0.25
Cl^-	$CF_3CO_2CH_3$	0.045	0.021
CN^-	$CF_3CO_2CH_3$	0.03	0.01
Br^-	$CF_3CO_2CH_3$	~0.005	~0.003

is the gas-phase equivalent of the S_N2 reaction so extensively studied in solution. Leider and Brauman[123,124] in a study of the reaction of Cl^- with *cis-* and *trans-* 4-bromocyclo-hexanol in the gas phase, have shown that the reaction occurs with inversion of configuration as expected for a true S_N2 displacement reaction. Displacement reactions generally are observed if they are exothermic and if there is no facile proton transfer process from RY to X^-. For example, the OH^- ion reacts with CH_3CN by proton abstraction rather than by displacement of CN^-.[68]

Table 13 presents a summary of rate constants for some simple displacement reactions taken from the work of Olmstead and Brauman;[125] also given in the table are the reaction efficiencies defined as k_{react}/k_{collis}, with k_{collis} calculated by the ADO theory. It should be noted that other measurements[126-128] on the same reactions in some cases report rate constants differing by up to a factor of two. A wide range of reaction efficiencies is observed; these do not correlate with the reaction thermochemistry but rather depend in detail on the specific nucleophile X^- and leaving group Y^-. The results

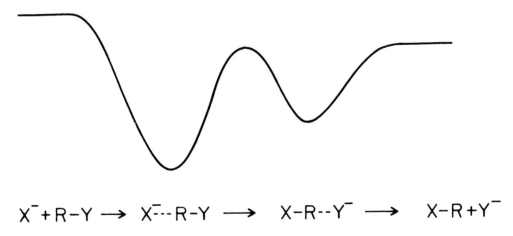

$$X^- + R-Y \longrightarrow X^{\overline{\cdots}}R-Y \longrightarrow X-R\cdots Y^- \longrightarrow X-R+Y^-$$

FIGURE 7.   Potential energy surface for displacement reaction.

have been rationalized by Brauman and colleagues[125,129] in terms of a potential surface for the reaction involving two minima as illustrated in Figure 7. The height of the barrier between the two minima strongly influences the rate of reaction. A similar potential energy surface probably should be invoked when considering proton transfer reactions involving either positive or negative ions.

Displacement reactions are of minimal utility in chemical ionization mass spectrometry since they provide no molecular weight information and minimal structural information; indeed, in unfavorable cases they may be a distinct disadvantage.

D. Proton Transfer Reactions

The proton transfer reaction

$$X^- + YH \rightarrow XH + Y^- \tag{41}$$

will be exothermic if the gas-phase acidity of YH is greater than the gas-phase acidity of XH. The gas-phase acidity scale is defined (Section VI.D.) in terms of the $\Delta H°$ for the reaction

$$AH \rightarrow A^- + H^+ \quad \Delta H° = \Delta H_{acid} \tag{42}$$

Since $\Delta H^°_{42}$ is positive, the acidity is greater the smaller the ΔH_{acid}. Alternatively $\Delta H^°_{42}$ represents the proton affinity of A^- and an equivalent statement is that Reaction 41 will be exothermic if the proton affinity of X^- is greater than the proton affinity of Y^-.

Proton transfer reactions have been studied extensively under equilibrium conditions to establish relative gas-phase acidities,[18,130] however, there have been fewer kinetic studies. Figure 8 summarizes results obtained by Bohme[131] for reaction of H^-, NH_2^-, OH^-, and CH_3O^- with a variety of simple molecules plotted as the reaction efficiency (k_{exp}/k_{collis}) vs. $\Delta H°$ for the reaction, where k_{collis} is the collision rate constant calculated from ADO theory. As shown, these reactions show unit efficiency provided they are at least 10 kcal mol^{-1} exothermic. The reaction efficiencies tend to drop off as the reaction approaches thermoneutrality and become low for endothermic reactions. For reactions involving delocalized reactant ions, such as $CH_3COCH_2^-$, rather low reaction efficiencies are observed.[132] It appears from the results to date that proton abstraction reactions by localized anions will be relatively efficient provided they are exothermic.

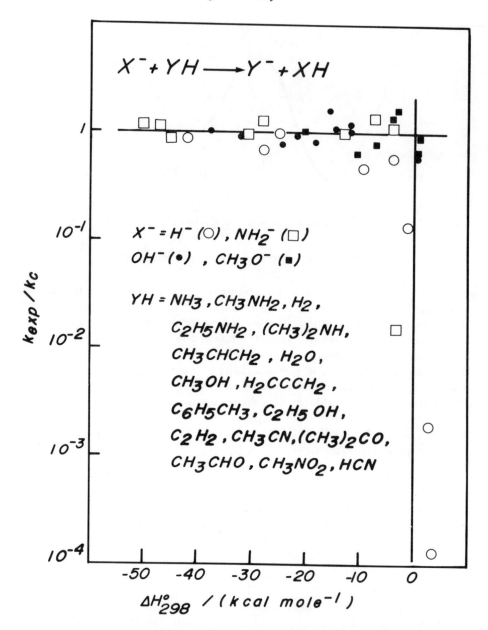

FIGURE 8.   Efficiency of proton abstraction as a function of ΔH°_{react}. (From Bohme, D. K., *Trans. R. Soc. Can.*, in press. With permission.)

Such proton transfer reactions are of importance in chemical ionization mass spectrometry not only because of their high efficiency but also because they lead to formation of $[M-H]^-$ ions which provide molecular weight information. The anions OH^- and CH_3O^-, which have found considerable use in chemical ionization, appear to react exclusively by proton transfer provided such proton transfer is exothermic. By contrast the $O^{\overline{\cdot}}$ ion reacts with organic molecules in part by proton transfer but also reacts by $H_2^{\overline{\cdot}}$ and H-atom abstraction from the organic substrate.[133-136] For example, $O^{\overline{\cdot}}$ reacts with acetone by H-atom abstraction (41%), H^+ transfer (7%) and $H_2^{\overline{\cdot}}$ transfer (52%) with a total reaction rate constant of 1.7×10^{-9} cm³ molecule⁻¹ sec⁻¹.[133]

Table 14
CHARGE EXCHANGE REACTIONS OF
NEGATIVE IONS

Reaction	$k \times 10^{10}$ cm^3 molecule^{-1} sec^{-1}	Ref.
$O^- + NO_2 \rightarrow NO_2^- + O$	12	122
$O_2^- + NO_2 \rightarrow NO_2^- + O_2$	20	137
$O_2^- + O_3 \rightarrow O_3^- + O_2$	3	138
$NO^- + O_2 \rightarrow O_2^- + NO$	9	138
$O_3^- + NO_2 \rightarrow NO_2^- + O_3$	7	137

Table 15
ASSOCIATION REACTIONS OF NEGATIVE IONS

Reaction	k (cm^6 molecule^{-2} sec^{-1})	Ref.
$O^- + 2O_2 \rightarrow O_3^- + O_2$	9×10^{-31}	140
$O^- + 2CO_2 \rightarrow CO_3^- + CO_2$	8×10^{-29}	141
$O_2^- + 2CO_2 \rightarrow CO_4^- + CO_2$	9×10^{-30}	141
$O_2^- + CO_2 + O_2 \rightarrow CO_4^- + O_2$	2.0×10^{-29}	141
$O_2^- + H_2O + O_2 \rightarrow O_2^-(H_2O) + O_2$	2.2×10^{-28}	142
$O_2^-(H_2O) + H_2O + O_2 \rightarrow O_2^-(H_2O)_2 + O_2$	6×10^{-28}	142

E. Charge Exchange Reactions

The charge exchange reaction

$$A^- + B \rightarrow B^- + A \qquad (43)$$

will be exothermic provided the electron affinity of B is greater than the electron affinity of A. The rate constants reported for a number of simple exothermic negative ion charge exchange reactions are summarized in Table 14. These limited results indicate that exothermic charge exchange reactions are 30 to 100% efficient; similar results might be expected for the more complex systems involved in chemical ionization. In addition to exothermic charge exchange reactions, endothermic charge exchange reactions have been studied as a function of ion kinetic energy; the energy threshold for the endothermic reaction is used to establish the difference in electron affinities of the reactants.[139]

F. Association Reactions

Third-order association or clustering reactions are as common for negative ions as for positive ions. Thus, anions, in the presence of polar molecules such as water, form solvated species. Although the equilibria involved in such solvation reactions have been studied extensively,[18,105] the kinetic parameters do not appear to have been extensively determined. Table 15 reports rate constants measured for a number of association reactions. The third-order rate constants are similar in magnitude to those observed for similar positive ion reactions (Table 9) including those for solvation reactions.

V. SENSITIVITY OF CHEMICAL IONIZATION

A question of considerable importance is the sensitivity (ion current of additive per unit additive pressure) of the chemical ionization method. As will be seen below, the sensitivity depends strongly on the rate constants discussed in the previous sections. Absolute sensitivities are difficult to estimate because of unknown and variable ion source extraction efficiencies and mass spectrometer transmissions. However, the sensitivity of chemical ionization relative to electron impact ionization on the same instrument can be derived. The following treatment for conventional CI and EI sources is based largely on the development by Field.[143]

The total ion current produced by electron impact, I_{EI}, is given by

$$I_{EI} = I_e \, Q \, \ell \, N \tag{44}$$

where I_e is the ionizing electron current, Q is the molecular ionization cross section, ℓ is the ionizing path length, and N is the concentration of molecules per cm^3. For chemical ionization the additive ion current, I_{CI}, is given by the total number of reagent ions produced times the fraction $(1 - \exp(-kNt))$ which react with the additive. Under usual CI conditions the ionizing electron beam is completely attenuated with the result that the total reagent ion current is given by the ionizing electron current, I_e, multiplied by the number, a, of ion pairs produced per electron. (Typically the energy required to produce an ion pair is ~ 30 eV[144] so that $a \approx 10$ for 300 eV electrons.) Thus

$$I_{CI} = aI_e[1 - \exp(-kNt)] \approx a \, I_e \, kNt \tag{45}$$

where k is the rate constant for reaction of the reagent ion(s) with the additive and t is the ion source residence time available for such reaction. For identical values of I_e and N we derive

$$\frac{I_{CI}}{I_{EI}} = \frac{akt}{Q\ell} \tag{46}$$

Substituting typical values $a = 10$, $k \approx 2 \times 10^{-9}$ cm^3 molecule^{-1} sec^{-1}, $t = 10^{-5}$ sec, $\ell = 1$ cm, and $Q = 2 \times 10^{-15}$ cm^2 molecule^{-1} [145] we obtain $I_{CI}/I_{EI} \approx 100$. This value represents the ratio of ion currents produced by equal electron beams in the ion source. Assuming the extraction efficiency from the ion source is proportional to the area of the ion exit slit the collected ion current ratio can be reduced by a factor of 10 to 20 due to the use of a smaller ion exit slit in the chemical ionization source. In addition, the electron beam entrance slit often is reduced in a CI source, thus reducing the value of I_e in CI relative to EI and further reducing the ratio. Nevertheless these approximate calculations do indicate that the collected ion currents in chemical ionization are at least as intense as those observed in electron impact instruments. In addition, the CI ion current may be concentrated in fewer ionic species.

The ion current produced in chemical ionization, and hence the sensitivity of the method, is directly dependent on k, the rate constant for the chemical ionization reaction. As the discussion in Sections III and IV of this chapter has shown, many ion/molecule reactions used as chemical ionization reactions are highly efficient and have rate constants in the range 1 to 4×10^{-9} cm^{-3} molecule^{-1} sec^{-1}; these CI systems will show approximately equal sensitivities. Inefficient ionization reactions with slower rate constants will show lower sensitivities. In particular, endothermic ionization reactions will have an activation energy at least equal to the endothermicity and will be correspondingly inefficient. For example, an activation energy of only 2 kcal mol^{-1} will decrease the rate of the ionization reaction by more than a factor of 10.

Finally it should be noted that electron attachment reactions can have rate constants up to 10^{-7} cm^3 molecule^{-1} sec^{-1} (Table 11) compared to a maximum of $\sim 4 \times 10^{-9}$ cm^3 molecule^{-1} sec^{-1} for ion/molecule reactions, the very high rate constants being a result of the high mobility of the electron. Thus chemical ionization which relies on electron capture reactions can in suitable cases have sensitivities which are up to a factor of 100 greater than those relying on ion/molecule reactions.[118,119] However, as the data in Table 11 show, the rate constants for electron attachment are very strongly structure-dependent and this method of ionization must be used carefully.

VI. THERMOCHEMICAL PROPERTIES OF GAS-PHASE IONS

In this section we review the thermochemical properties of gas-phase ions which are of relevance to chemical ionization mass spectrometry. Many of these thermochemical results have been derived from equilibrium studies of gas-phase ion/molecule reactions.

A. Gas-Phase Basicities: Proton Affinities

The fundamental concept of basicity, defined by Brønsted[146] as the tendency of a molecule B to accept a proton in the reaction

$$B + H^+ \rightarrow BH^+ \tag{47}$$

has long been of concern to chemists. By definition $-\Delta H^\circ_{47}$ is designated as the proton affinity (PA) of the base B. Although relative basicities in solution have been established for some time it is only relatively recently that gas-phase basicities, free of the complication of solvation, have been measured by mass spectrometric techniques. Early gas-phase measurements used reaction bracketing techniques[147-148] to establish the direction of exothermic proton transfer between base pairs, thus providing approximate relative proton affinities. In effect rapid proton transfer reactions were assumed to be exothermic and slow reactions were assumed to be endothermic. More recently quantitative measurements of equilibrium constants for reactions of the type

$$B_1 H^+ + B_2 \rightleftharpoons B_2 H^+ + B_1 \tag{48}$$

by ion cyclotron resonance techniques,[149] by high-pressure mass spectrometry,[150,151] and by flowing afterglow experiments[43,152] have permitted the determination of relative proton affinities to a precision of ± 0.2 kcal mol^{-1}. By the use of suitable reference compounds, whose proton affinities are known from other measurements, these relative values can be put on an absolute scale. Most of the equilibrium measurements have been made at a single temperature and thus yield ΔG° for Reaction 48. To establish ΔH°, the difference in proton affinities, ΔS° must be evaluated. It usually has been assumed that the only contribution to ΔS° arises from symmetry changes; this assumption is supported by the few measurements made of the temperature dependence of the equilibrium where ΔH° and ΔS° are evaluated experimentally.[150,151,153,154]

A number of extensive reviews of gas-phase proton transfer equilibria are available.[43,105,155,156] In these reviews the absolute values of proton affinities, but not the relative values, show some differences because the absolute assignment of the proton affinity scale is still in some doubt and the different authors have chosen different standards to establish the absolute scale. In the most recent review Aue and Bowers[156] have based the scale on PA(NH$_3$) = 205.0 kcal mol^{-1} and PA(i-C$_4$H$_8$) = 196.9 kcal mol^{-1}; this scale is adapted in the present tabulation of proton affinities of organic molecules. Tables 16 to 20 summarize a large body of proton affinity data; more complete tabulations are available elsewhere.[156]

A detailed evaluation of the proton affinity data and the variation of proton affinity with structure has been presented elsewhere[156] and need not be repeated here. For the present purposes it is sufficient to note that oxygen-containing compounds have proton affinities in the 180 to 205 kcal mol^{-1} range, sulfur-containing compounds in the 198

Table 16
PROTON AFFINITIES
OF SUBSTANCES
LESS BASIC THAN
H_2O

B	PA(B)[a] (kcal mol^{-1})
O_2	100.4
H_2	100.7
Kr	101.4
N_2	117.4
Xe	118.0
CO_2	128.6
CH_4	130.5
N_2O	137.0
CO	141.4

[a]   From Reference 70.

Table 17
PROTON AFFINITIES
OF ALKENES AND
ALKADIENES

Compounds	PA (kcal mol^{-1})[a]
	163.5
	184.9
	182.0
	196.9
	193
	201.8
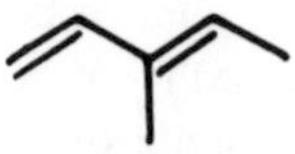	205.7
	200.0

[a]   Calculated from data in Reference 156.

to 209 kcal mol^{-1} range, and nitrogen-containing compounds in the 205 to 240 kcal mol^{-1} range. Benzenoid compounds have proton affinities in the 182 to 212 kcal mol^{-1} range depending on the substituent. Thus it is clear that BH$^+$ ions derived from any of the bases B listed in Table 16 as well as the $C_2H_5^+$ (protonated ethylene) ions also produced in high-pressure methane will protonate all organic bases in an exothermic reaction. On the other hand protonation of some organic compounds by CI reagent ions such as $C_4H_9^+$ (protonated isobutene) or NH_4^+ will be endothermic and thus extremely slow. In these cases other reaction modes, such as clustering, may become important.

Table 18
PROTON AFFINITIES
OF SUBSTITUTED
BENZENES

Substituent	PA (kcal mol^{-1})[a]
Cl	181.7
F	181.5
H	182.8
CH$_3$	191.2
C$_2$H$_5$	192.2
n-C$_3$H$_7$	191.0[b]
i-C$_3$H$_7$	191.4[b]
t-C$_4$H$_9$	191.6[b]
NO$_2$	193.8
OH	196.2
CN	196.3
CHO	200.3
OCH$_3$	200.6
NH$_2$	211.5

[a] From reference 153.
[b] From reference 157.

Table 19
PROTON AFFINITIES BETWEEN H$_2$O AND NH$_3$[a]

Alcohols and ethers		Aldehydes and ketones		Acids and esters		Thiols, sulfides, nitriles	
H$_2$O	173.0	HCHO	177.2	CF$_3$CO$_2$H	176.0	H$_2$S	176.6
CF$_3$CH$_2$OH	174.9	MeCHO	188.9	HCO$_2$H	182.8	MeSH	188.6
MeOH	184.9	EtCHO	191.4	MeCO$_2$H	190.7	EtSH	192.0
EtOH	190.3	n-PrCHO	193.6	EtCO$_2$H	193.4	i-PrSH	194.7
n-PrOH	191.4	n-BuCHO	193.3	HCO$_2$Me	190.4	Me$_2$S	200.7
t-BuOH	195.0	Me$_2$O	197.2	HCO$_2$Et	194.2	Et$_2$S	205.6
Me$_2$O	193.1	MeCOEt	199.4	HCO$_2$n-Pr	195.2	i-Pr$_2$S	209.3
Et$_2$O	200.4			MeCO$_2$Me	198.3	HCN	178.9
n-Pr$_2$O	202.9			MeCO$_2$Et	201.3	MeCN	190.9
n-Bu$_2$O	203.9			MeCO$_2$n-Pr	202.0	EtCN	192.8
THF	199.6					n-PrCN	193.8
THP	200.7						

[a] PAs (in kcal mol^{-1}) from Reference 156.

The proton affinity data in the tables refer to protonation at the most basic site of the molecule and for these molecules with more than one basic site the question arises as to the site of protonation. For simple carboxylic acids and esters, correlations of the PA data with O(1S) ionization energies have established[158-160] that the compounds are protonated at the carbonyl oxygen; the proton affinities of the singly-bonded oxygens are 18 to 25 kcal mol^{-1} lower.[160] Similar results indicate that amides are preferentially protonated at the oxygen.[161] For substituted benzenes protonation of the aromatic ring occurs except for those with NO$_2$, CHO, and CN substituents where protonation occurs at the substituent.[153] These results refer to protonation of the substrate under conditions of thermodynamic control; by contrast, most chemical ionization protonation reactions will be under kinetic control and the kinetically favored site

Table 20
PROTON AFFINITIES GREATER THAN NH₃ᵃ

Amines		α,ω-Diamines		Pyridines		Anilines	
NH_3	205.0	$H_2N(CH_2)_2NH_2$	226.5	H	220.4	H	211.5
$MeNH_2$	214.1	$H_2N(CH_2)_3NH_2$	232.7	2-Me	223.7	m-Me	214.3
$EtNH_2$	217.1	$H_2N(CH_2)_4NH_2$	238.1	3-Me	222.8	p-Me	214.3
n-Pr NH_2	218.5	$H_2N(CH_2)_5NH_2$	235.8	4-Me	223.7	m-MeO	216.7
n-Bu NH_2	219.0	$Me_2N(CH_2)_4NMe_2$	245.9	2,4-Me₂	226.9	p-MeO	215.0
Me_2NH	220.5	$Me_2N(CH_2)_6NMe_2$	249.0	3-CN	209.5	m-Cl	208.6
Me_3N	224.3			3-F	214.8	p-Cl	209.6
Et_2NH	225.1			3-Cl	215.7	m-F	208.2
Et_3N	231.2			3-MeO	222.5	C_6H_5NHMe	218.2
n-Pr₂NH	227.4			3-NH₂	221.0	$C_6H_5NMe_2$	223.8
n-Pr₃N	233.4			4-NO₂	209.5		

ᵃ   PAs (in kcal mol⁻¹) from Reference 156.

of protonation may differ from the thermodynamically favored site. Consequently it is not safe to assume that in chemical ionization protonation occurs only at the thermodynamically favored site. For example, it has been shown[162] that in the chemical ionization of halobenzenes significant protonation occurs at the halogen even though protonation of the aromatic ring is thermodynamically favored. In many cases the proton may interchange between the available basic sites in the molecule; this has been shown to be the case in the protonation of amino acids.[163]

B. Hydride Ion Affinities
The hydride ion affinity (HIA) of a gaseous cation is defined as $-\Delta H°$ for the reaction

$$R^+ + H^- \rightarrow RH - \Delta H° = HIA(R^+) \tag{49}$$

Since $\Delta H_f°(H^-) = 33.4$ kcal mol⁻¹,[164] hydride ion affinities can be calculated if $\Delta H_f°(R^+)$ and $\Delta H_f°(RH)$ are known. Relative hydride ion affinities also can be obtained in favorable cases from studies[166,167] of the equilibrium

$$R_1^+ + R_2H \rightleftharpoons R_2^+ + R_1H \tag{50}$$

A partial list of hydride ion affinities is given in Table 21; a more complete listing can be found elsewhere.[156]

The hydride ion affinity scale is probably the most useful and convenient basis for the comparison of relative carbocation stabilities. Of more relevance in the present context is the ordering of cations with respect to the ability to abstract H^- in chemical ionization reactions. The hydride ion abstraction Equation 50 will be exothermic provided HIA (R_1^+)>HIA (R_2^+). Accordingly CH_3^+, CF_3^+, and H_3^+, with high hydride ion affinities, should be capable of abstracting H^- from most organic substrates. The CH_5^+ and $C_2H_5^+$ ions are slightly weaker H^--abstracting reagent ions while NO^+ is a relatively weak H^--abstracting reagent. The t-$C_4H_9^+$ ion is quite a weak H^--abstracting reagent. Of course a number of these species are Brønsted acids which also can react by proton transfer; the competition between these two reaction modes is discussed in more detail in Chapter 4.

C. Electron Affinities
A fundamental property of a negative ion is the lowest energy required to remove an electron. This energy is defined as the electron affinity (EA) of the anion. For the reaction

Table 21
HYDRIDE ION AFFINITIES OF GASEOUS IONS[a]

R+	$\Delta H^\circ{}_f (R^+)$[b]	$\Delta H^\circ_f (RH)$[c]	HIA (R+)
CH_3^+	261	−17.9	312
H_3^+	266.5[d]	0(2H_2)	300
CF_3^+	99.3	−166.3	299
$C_2H_5^+$	219	−20.2	272
CH_5^+	218.8[d]	−17.9($H_2 + CH_4$)	270
NO^+	237.5[e]	24[e]	246
$n\text{-}C_3H_7^+$	208	−24.8	266
$i\text{-}C_3H_7^+$	192	−24.8	250
$n\text{-}C_4H_9^+$	201	−30.2	264
$i\text{-}C_4H_9^+$	199	−32.2	264
$s\text{-}C_4H_9^+$	183	−30.2	246
$t\text{-}C_4H_9^+$	166	−32.2	231
$CH_2{=}CH{-}CH_2^+$	226	4.9	254
$CH_2{=}CH{-}\overset{+}{C}HCH_3$	204	−0.03	237
$C_6H_5CH_2^+$	212	12.0	233
$CH_3CH{=}OH^+$	138	−56	228
$(CH_3)_2C{=}OH^+$	118	−65	216

[a] All data in kcal mol^{-1}.
[b] From Reference 156, unless otherwise indicated.
[c] From Reference 167, unless otherwise indicated.
[d] From Reference 70.
[e] From Reference 168.

$$X^- \rightarrow X + e \qquad \Delta H^\circ = EA(X^-) \tag{51}$$

ΔH° = EA provided X^- and X are in their ground rotational, vibrational, and electronic states and provided the electron has zero kinetic and potential energy. If one wishes to focus on the neutral species X, ΔH°_{52} can be defined as the affinity of X for an electron; this is an entirely equivalent definition.

There are two general types of experiments that have been employed successfully in determining absolute electron affinities. In optical methods[169,170] photons interact with negative ions to remove the least tightly bound electron. This method has been employed either by using a fixed frequency light source and measuring the energy of the ejected electron or by varying the frequency of the light source and determining the frequency-dependence of the photodetachment cross section. The second type of experiment consists of studying the endothermic charge exchange reaction

$$X^- + Y \rightarrow Y^- + X \tag{52}$$

as a function of the translational energy of X^-.[139,171] From the observed translational energy threshold the endothermicity can be estimated and the electron affinity of Y^- determined if the electron affinity of X^- is known. This method gives only lower estimates for EA(Y^-) since Y^- is not necessarily formed in its ground state. The methods of determining electron affinities and the results obtained therefrom have been reviewed recently.[172]

Table 22 presents a listing of selected electron affinities; a more complete listing, including many enolate anions, can be found elsewhere.[172] There have been few measurements of the electron affinities of molecular anions. Benzene has a negative electron affinity but hexafluorobenzene and nitrobenzene have substantial positive electron affinities. Similarly the polycyclic compounds, azulene and fluoranthene, form

Table 22
TABLE OF ELECTRON
AFFINITIES

X^-	EA(X^-) (kcal mol^{-1})	Ref.
H^-	17.4	173
O^-	33.7	173
F^-	78.4	173
Cl^-	83.4	173
Br^-	77.6	173
NH_2^-	18.0	174
OH^-	42.2	174
O_2^-	10.1	174
Cl_2^-	53.5	175
F_2^-	71.0	176
CH_3O^-	36.2	177
CH_3S^-	43.4	177
$C_3H_5^-$	12.7	178
$c\text{-}C_5H_5^-$	41.2	179
$CH_3COCH_2^-$	41.3	180
$C_6H_6^-$	<0	181
$C_6F_6^-$	>41.5	182
$C_6F_5CF_3^-$	>39.2	182
$C_6H_5NO_2^-$	>16.1	182
Azulene ($C_{10}H_8^-$)	14.3	183
Fluoranthene ($C_{16}H_{10}^-$)	20.8	184

stable negative molecular anions. The usual negative ion chemical ionization reagent ions O^-, Cl^-, OH^-, and CH_3O^- have large electron affinities and are unlikely to undergo exothermic charge exchange reactions with most organic molecules. On the other hand O_2^- has a low electron affinity and may well react by charge exchange.

D. Gas-Phase Acidities: Proton Affinities of Anions

The gas-phase acidity scale is defined in terms of the $\Delta H°$ for the gas-phase reaction

$$XH \rightarrow X^- + H^+, \Delta H° = \Delta H°_{acid} \qquad (53)$$

Since Reaction 53 is invariably endothermic the gas-phase acidity decreases as $\Delta H°_{acid}$ increases. Alternatively $\Delta H°_{53}$ can be considered as the proton affinity of the anion X^-. In some cases absolute acidities can be evaluated from bond dissociation energies and electron affinities using the following thermochemical cycle:

$$XH \rightarrow X + H \qquad \Delta H° = D(X - H) \qquad (54)$$

$$X + H \rightarrow X^- + H^+ \qquad \Delta H° = IE(H) - EA(X) \qquad (55)$$

$$XH \rightarrow X^- + H^+ \qquad \Delta H°_{acid} = D(X - H) + IE(H) - EA(X) \quad (56)$$

Since the ionization energy of H is known (IE(H) = 313.6 kcal mol^{-1}),[185] the absolute acidity of HX can be evaluated if the bond dissociation energy, D(H$-$X), and the electron affinity, EA(X), are known. Alternatively, relative acidities can be established from measurements of the equilibrium constants for the proton-transfer reaction

Table 23

ΔH°_{acid} FOR SIMPLE
BRØNSTED ACIDS[a]

$HX \rightleftharpoons H^+ + X^-$

HX	ΔH°_{acid} (kcal mol^{-1})
H_2	400.4
NH_3	399.6
H_2O	390.8
OH	382.1
CH_3OH	379.2
HF	371.5
HCN	353.1
HO_2	350.6
HCl	333.3

[a] Data from Reference 130.

Table 24

ΔH°_{acid} FOR ORGANIC COMPOUNDS[a]

Alcohols		Phenols		Carboxylic acids		Carbonyl compounds	
MeOH	379.2	H	349.8	$MeCO_2H$	348.5	CH_3CHO	366.4
EtOH	376.1	$m\text{-}CH_3$	350.3	$EtCO_2H$	347.3	$MeCH_2CHO$	365.9
n-PrOH	374.7	$m\text{-}CH_3O$	348.3	HCO_2H	345.2	$(CH_3)_2CO$	368.8
i-PrOH	374.1	$m\text{-}NH_2$	350.7	$PhCO_2H$	338.8	$PhCOCH_3$	363.2
i-BuOH	373.4	m-F	344.0	FCH_2CO_2H	337.6	$PhCH_2COCH_3$	352.5
t-BuOH	373.3	m-Cl	341.9	$ClCH_2CO_2H$	335.4	CH_3CONMe_2	373.5
$PhCH_2OH$	369.6	m-CN	335.5	CF_3CO_2H	322.7	CH_3COCF_3	350.3
F_2CHCH_2OH	367.0	$m\text{-}NO_2$	334.1	$MeCO_2Me$	371.0	$PhCOCH_2COMe$	339.9
F_3CCH_2OH	364.4						

Thiols		Acetylenes		Nitriles and nitro compounds		Hydrocarbons		Anilines	
MeSH	359.0	$HC\equiv CH$	375.4	CH_3CN	372.2	$c\text{-}C_5H_6$	356.1	H	367.1
EtSH	357.4	$MeC\equiv CH$	379.6	$MeCH_2CN$	373.7	$c\text{-}C_7H_8$	373.9	p-MeO	367.9
n-PrSH	356.4	$nPrC\equiv CH$	378.3	$PhCH_2CN$	353.3	$PhCH_3$	379.0	m-Me	368.2
i-PrSH	355.6	$t\text{-}BuC\equiv CH$	376.6	CH_3NO_2	358.7	$PhCH_2Me$	378.3	m-MeS	364.0
t-BuSH	354.7	$PhC\equiv CH$	370.3	$MeCH_2NO_2$	358.1	Ph_2CH_2	364.5	m-F	363.1
						Fluorene	353.3	m-Cl	361.1
						$CH_3CH=CH_2$	390.8	$m\text{-}CF_3$	359.1

[a] Data in kcal mol^{-1} From Reference 130.

$$X_1^- + X_2H \rightleftharpoons X_2^- + X_1H \qquad (57)$$

These relative acidities can be put on an absolute scale by including in the measurements an acid whose acidity has been derived thermochemically. Equilibrium measurements at a single temperature lead directly to ΔG° values; to derive ΔH° values the standard entropy change must be evaluated. The entropy change normally has been derived by assuming that the only significant contributions to ΔS° are rotational entropy changes arising from symmetry changes and the freezing out or loss of internal rotations.[130] Equilibrium measurements of the proton transfer reactions of negative ions have been made by high-pressure mass spectrometry,[186,187] by ICR spectroscopy,[188] and by the flowing afterglow method.[189]

Table 23 summarizes the gas-phase acidities of Brønsted acids, XH, where the anion X^- has been, or could be used as a proton abstraction chemical ionization reagent ion. Table 24 presents a selection of acidities for a variety of organic molecules; a more complete tabulation will be found elsewhere,[130] as well as a discussion of structural effects on gas-phase acidities.[130,190] The compounds in Table 23 have been listed in order of increasing acidity or decreasing proton affinity of the corresponding anion. The larger ΔH°_{acid} (XH) the greater is the probability that the proton abstraction reaction of X^- will be exothermic. Obviously anions such as NH_2^-, OH^-, or $O^{\overline{.}}$ should be capable of exothermically abstracting a proton from a wide variety of organic molecules. On the other hand $O_2^{\overline{.}}$ and Cl^- are much weaker reagents and will show a limited capability for proton abstraction.

REFERENCES

1. Dempster, A. J., The ionization and dissociation of hydrogen molecules and the formation of H_3, *Phil. Mag.*, 31, 438, 1916.
2. Smyth, H. D., Primary and secondary products of ionization in hydrogen, *Phys. Rev.*, 25, 452, 1925.
3. Hogness, T. R. and Lunn, E. G., The ionization of hydrogen by electron impact as interpreted by positive ray analysis, *Phys. Rev.*, 26, 44, 1925.
4. Hogness, T. R. and Harkness, R. W., The ionization processes of iodine interpreted by the mass spectrograph, *Phys. Rev.*, 32, 784, 1928.
5. Smyth, H. D., Products and processes of ionization by low speed electrons, *Rev. Mod. Phys.*, 3, 347, 1931.
6. Thompson, J. J., *Rays of Positive Electricity*, Longmans Green, London, 1931.
7. Tal'roze, V. L. and Lyubimova, A. K., Secondary processes in a mass spectrometer ion source, *Dokl. Akad. Nauk. SSSR.*, 86, 969, 1952.
8. Stevenson, D. P. and Schissler, D. O., Rate of the gaseous reaction $X^+ + YH \rightarrow XH^+ + Y$, *J. Chem. Phys.*, 23, 1353, 1955.
9. Field, F. H., Franklin, J. L., and Lampe, F. W., Reactions of gaseous ions. I. Methane and ethylene, *J. Am. Chem. Soc.*, 79, 2419, 1957.
10. Lampe, F. W., Franklin, J. L., and Field, F. H., Kinetics of the reactions of ions with molecules, in *Progress in Reaction Kinetics*, Porter, G., Ed., Pergamon Press, Elmsford, N.Y., 1961, 69.
11. Tal'roze, V. L., Ion-molecule reactions in gases, *Pure Appl. Chem.*, 5, 455, 1962.
12. Stevenson, D. P., Ion-molecule reactions, in *Mass Spectrometry*, McDowell, C. A., Ed., McGraw-Hill, New York, 1963.
13. Melton, C. E., Ion-molecule reactions, in *Mass Spectrometry of Organic Ions*, McLafferty, F. W., Ed., Academic Press, New York, 1963.
14. Henchman, M. J., Ion-molecule reactions and reactions in crossed molecular beams, *Ann. Rep. Chem. Soc.*, 62, 39, 1965.
15. Tal'roze, V. L. and Karachevtsev, G. V., Ion-molecule reactions, *Adv. Mass Spectrom.*, 3, 211, 1966.
16. Futrell, J. H. and Tiernan, T. O., Ion-molecule reactions in *Fundamental Processes in Radiation Chemistry*, Ausloos, P., Ed., John Wiley & Sons, New York, 1968.
17. Ferguson, E. E., Ion-molecule reactions, *Ann. Rev. Phys. Chem.*, 26, 17, 1975.
18. Kebarle, P., Ion thermochemistry and solvation from gas phase ion equilibria, *Ann. Rev. Phys. Chem.*, 28, 445, 1977.
19. Jennings, K. R., Recent developments in the study of ion-molecule reactions, *Adv. Mass Spectrom.*, 7, 209, 1978.
20. Durup, J., *Les Reactions entre Ions Positif et Molecules en Phase Gaseuse*, Gauthier-Villars, Paris, 1960.
21. Ausloos, P., Ed., *Ion Molecule Reactions in the Gas Phase*, Adv. Chem. Ser. 58, American Chemical Society, Washington, D.C., 1966.
22. McDaniel, E. W., Cermak, V., Dalgarno, A., Ferguson, E. E., and Friedman, L., *Ion-Molecule Reactions*, Interscience, New York, 1970.
23. Franklin, J. L., Ed., *Ion-Molecule Reactions*, Plenum Press, New York, 1972.
24. Ausloos, P., Ed., *Interactions Between Ions and Molecules*, Plenum Press, New York, 1975.
25. Lias, S. G. and Ausloos, P., *Ion-Molecule Reactions. Their Role in Radiation Chemistry*, American Chemical Society, Washington, D.C., 1975.
26. Ausloos, P., Ed., *Kinetics of Ion-Molecule Reactions*, Plenum Press, New York, 1979.
27. Bowers, M. T., Ed., *Gas Phase Ion Chemistry*, Academic Press, New York, 1979.
28. Eyring, H., Hirschfelder, J. O., and Taylor, H. S., The theoretical treatment of chemical reactions produced by ionization processes. I. The ortho-para hydrogen conversion by alpha-particles, *J. Chem. Phys.*, 4, 479, 1936.
29. Vogt, E. and Wannier, G. H., Scattering of ions by polarization forces, *Phys. Rev.*, 95, 1190, 1954.
30. Gioumousis, G. and Stevenson, D. P., Reactions of gaseous ions with gaseous molecules. II. Theory, *J. Chem. Phys.*, 29, 294, 1958.
31. Langevin, P. M., Une formule fondamentale de theorie cinetique, *Ann. Chim. Phys.*, 5, 245, 1905.
32. McDaniel, E. W., *Collision Phenomena in Ionized Gases*, John Wiley & Sons, New York, 1964.
33. Henglein, A., Kinematics of ion-molecule reactions, in *Molecular Beams and Reaction Kinetics*, Schlier, C., Ed., Academic Press, New York, 1970.
34. Su, T. and Bowers, M. T., Classical ion-molecule collision theory, in *Gas Phase Ion Chemistry*, Bowers, M. T., Ed., Academic Press, New York, 1979.
35. Moran, T. F. and Hamill, W. H., Cross-sections of ion permanent dipole reactions by mass spectrometry, *J. Chem. Phys.*, 39, 1413, 1963.

36. Theard, L. P. and Hamill, W. H., The energy dependence of cross-sections of some ion-molecule reactions, *J. Am. Chem. Soc.*, 84, 1134, 1962.

37. Gupta, S. K., Jones, E. G., Harrison, A. G., and Myher, J. J., Reactions of thermal energy ions. VI. Hydrogen transfer ion-molecule reactions involving polar molecules, *Can. J. Chem.*, 45, 3107, 1967.

38. Bowers, M. T. and Laudenslager, J. B., Mechanism of charge transfer reactions: reactions of rare gas ions with the trans-, cis-, and 1,1-difluoroethylene geometric isomers, *J. Chem. Phys.*, 56, 4711, 1972.

39. Su, T. and Bowers, M. T., Theory of ion-polar molecule collisions. Comparison with experimental charge transfer reactions of rare gas ions to geometric isomers of difluorobenzene and dichloroethylene, *J. Chem. Phys.*, 58, 3027, 1973.

40. Su, T. and Bowers, M. T., Ion-polar molecule collisions: the effect of ion size on ion-polar molecule rate constants: the parameterization of the average dipole orientation theory, *Int. J. Mass Spectrom. Ion Phys.*, 12, 347, 1973.

41. Bass, L., Su, T., and Bowers, M. T., Ion-polar molecule collisions. A modification of the average dipole orientation theory: the $\overline{\cos\theta}$ model, *Chem. Phys. Lett.*, 34, 119, 1975.

42. Su, T. and Bowers, M. T., Parameterization of the average dipole orientation theory: temperature dependence, *Int. J. Mass Spectrom. Ion Phys.*, 17, 211, 1975.

43. Bohme, D. K., The kinetics and energetics of proton transfer, in *Interactions Between Ions and Molecules*, Ausloos, P., Ed., Plenum Press, New York, 1975.

44. Harrison, A. G., Lin, P.-H., and Tsang, C. W., Proton transfer reactions by trapped ion mass spectrometry, *Int. J. Mass Spectrom. Ion Phys.*, 19, 23, 1976.

45. Huntress, W. T., Mosesman, M. M., and Elleman, D. D., Relative rates and their dependence on kinetic energy for ion-molecule reactions in ammonia, *J. Chem. Phys.*, 54, 843, 1971.

46. Bowers, M. T., Su, T., and Anicich, V. G., Theory of ion-polar molecule collisions, Kinetic energy dependence of ion-polar molecule reactions: $CH_3OH^+ + CH_3OH \rightarrow CH_3OH_2^+ + CH_3O$, *J. Chem. Phys.*, 58, 5175, 1973.

47. Su, T., Su, E. C. F., and Bowers, M. T., Ion-polar molecule collisions. Conservation of angular momentum in the average dipole orientation theory. The AADO theory, *J. Chem. Phys.*, 69, 2243, 1978.

48. Su, T. and Bowers, M. T., Ion-polar molecular collisions: the average quadrupole orientation theory, *Int. J. Mass Spectrom. Ion Phys.*, 17, 309, 1975.

49. Lindholm, E., Charge exchange and ion-molecule reactions observed in double mass spectrometers, in *Ion-Molecule Reactions in the Gas Phase*, Ausloos, P., Ed., American Chemical Society, Washington, D.C., 1966.

50. Lindholm, E., Mass spectra and appearance potentials studied by use of charge exchange in a tandem mass spectrometer, in *Ion-Molecule Reactions*, Franklin, J. L., Ed., Plenum Press, New York, 1972.

51. Herman, J. A., Li, Y.-H., and Harrison, A. G., Energy dependence of the fragmentation of some isomeric $C_6H_{12}^+$ ions, *Org. Mass Spectrom.*, 17, 143, 1982.

52. Bowers, M. T. and Elleman, D. D., Thermal energy charge transfer reactions of rare gas ions to methane, ethane, propane, and silane. The importance of Franck-Condon factors, *Chem. Phys. Lett.*, 16, 486, 1972.

53. Ausloos, P., Eyler, J. R., and Lias, S. G., Thermal energy charge transfer reactions involving CH_4 and SiH_4. Lack of evidence for non-spiralling collisions, *Chem. Phys. Lett.*, 30, 21, 1975.

54. Li, Y.-H. and Harrison, A. G., Bimolecular reactions of trapped ions. XIII. Charge transfer from Ar^+, Kr^+, and N_2^+ to methane, *Int. J. Mass Spectrom. Ion Phys.*, 28, 289, 1978.

55. Li, Y.-H. and Harrison, A. G., unpublished results.

56. Gauglhofer, J. and Kevan, L., Thermal energy charge transfer reactions of various projectiles to methane. Resonance effects in non-orbiting collisions, *Chem. Phys. Lett.*, 16, 492, 1972.

57. Huntress, W. T. and Pinnizotto, R. F., Product distributions and rate constants for ion-molecule reactions in water, hydrogen sulfide, ammonia, and methane, *J. Chem. Phys.*, 59, 4742, 1973.

58. Hellner, L. and Sieck, L. W., High pressure photoionization mass spectrometry. Effect of internal energy and density on the ion-molecule reactions occurring in methyl, dimethyl, and trimethyl amine, *Int. J. Chem. Kin.*, 5, 177, 1973.

59. Herod, A. A., Harrison, A. G., and McAskill, N. A., Ion-molecule reactions in methyl fluoride and methyl chloride, *Can. J. Chem.*, 49, 2217, 1971.

60. Solka, B. H. and Harrison, A. G., Bimolecular reactions of trapped ions. X. Reactions in methyl mercaptan and mixtures with methane, *Int. J. Mass Spectrom. Ion Phys.*, 14, 295, 1974.

61. Abramson, F. P. and Futrell, J. H., Ion-molecule reactions of methane, *J. Chem. Phys.*, 45, 1925, 1966.

62. Harrison, A. G., Heslin, E. J., and Blair, A. S., Bimolecular reactions of trapped ions. IV. Ion-molecule reactions in ethane and mixtures with C_2H_2 and CD_4, *J. Am. Chem. Soc.*, 94, 2935, 1972.

63. Solka, B. H., Lau, A. Y. K., and Harrison, A. G., Bimolecular reactions of trapped ions. VIII. Reactions in propane and propane-methane mixtures, *Can. J. Chem.*, 52, 1798, 1974.

64. Solka, B. H. and Harrison, A. G., Bimolecular reactions of trapped ions. IX. Effect of method of preparation on rate of reaction of $C_2H_5^+$, *Int. J. Mass Spectrom. Ion Phys.*, 14, 125, 1974.

65. Su, T. and Bowers, M. T., Ion-polar molecule collisions. The effect of molecular size on ion-polar molecule rate constants, *J. Am. Chem. Soc.*, 95, 7609, 1973.

66. Su, T. and Bowers, M. T., Ion-polar molecule collisions. Proton transfer reactions of $C_4H_9^+$ ions with NH_3, CH_3NH_2, $(C_2H_5)_2NH$, and $(CH_3)_3N$, *J. Am. Chem. Soc.*, 95, 7611, 1973.

67. Hemsworth, R. S., Payzant, J. D., Schiff, H. I., and Bohme, D. K., Rate constants at 297°K for proton transfer reactions with NH_3. Comparisons with classical theories and exothermicities, *Chem. Phys. Lett.*, 26, 417, 1974.

68. Mackay, G. I., Betkowski, L. D., Payzant, J. D., Schiff, H. I., and Bohme, D. K., Rate constants at 279K for proton transfer reactions with HCN and CH_3 CN. Comparisons with classical theories and exothermicity, *J. Phys. Chem.*, 80, 2919, 1976.

69. Solka, B. H. and Harrison, A. G., Bimolecular reactions of trapped ions. XI. Rate and equilibria in proton transfer reactions of $CH_3SH_2^+$, *Int. J. Mass Spectrom. Ion Phys.*, 17, 379, 1975.

70. Bohme, D. K., Mackay, G. I., and Schiff, H. I., Determination of proton affinities from the kinetics of proton transfer reactions. VII. The proton affinities of O_2, H_2, Kr, O, N_2, Xe, CO_2, CH_4, N_2O, and CO, *J. Chem. Phys.*, 73, 4976, 1980.

71. Field, F. H. and Lampe, F. W., Reactions of gaseous ions. VI. Hydride ion transfer reactions, *J. Am. Chem. Soc.*, 80, 5587, 1958.

72. Ausloos, P. and Lias, S. G., Structure and reactivity of hydrocarbon ions, in *Ion-Molecule Reactions*, Franklin, J. L., Ed., Plenum Press, New York, 1972.

73. Lias, S. G., Ion-molecule reactions in radiation chemistry, in *Interactions Between Ions and Molecules*, Ausloos, P., Ed., Plenum Press, New York, 1975.

74. Searles, S. K. and Sieck, L. W., High pressure photoionization mass spectrometry. III. Reactions of NO^+ $(X^1\Sigma^+)$ with C_3-C_7 hydrocarbons at thermal kinetic energies, *J. Chem. Phys.*, 53, 794, 1970.

75. Lias, S. G., Eyler, J. R., and Ausloos, P., Hydride ion transfer reactions involving saturated hydrocarbons and CCl_3^+, CCl_2H^+, CCl_2F^+, CF_2Cl^+, CF_2H^+, CF_3^+, NO^+, $C_2H_5^+$, sec-$C_3H_7^+$, and t-$C_4H_9^+$, *Int. J. Mass Spectrom. Ion Phys.*, 19, 219, 1976.

76. Ausloos, P. and Lias, S. G., Carbonium ions in radiation chemistry. Reactions of t-butyl ions with hydrocarbons, *J. Am. Chem. Soc.*, 92, 5037, 1970.

77. Lias, S. G., Rebbert, R. E., and Ausloos, P., Gas phase radiolysis of hydrocarbon mixtures: determination of the charge recombination rate coefficient and absolute rate constants of ion-molecule reactions of the t-butyl ion through a competitive kinetic method, *J. Chem. Phys.*, 57, 2080, 1972.

78. Hunt, D. F. and Harvey, J. M., Nitric oxide chemical ionization mass spectra of alkanes, *Anal. Chem.*, 47, 1965, 1975.

79. Hunt, D. F. and Ryan, J. F., Chemical ionization mass spectrometry studies. Nitric oxide as a reagent gas, *J. Chem. Soc. Chem. Commun.*, 620, 1972.

80. Doepker, R. D. and Ausloos, P., Gas-phase radiolysis of cyclopentane. Relative rates of H_2^--transfer reactions from various hydrocarbons to $C_3H_6^+$, *J. Chem. Phys.*, 44, 1951, 1965.

81. Gordon, R., Doepker, R., and Ausloos, P., Photoionization of propylene at 1236 Å. Reactions of $C_3D_6^+$ with added alkanes, *J. Chem. Phys.*, 44, 3733, 1965.

82. Sieck, L. W. and Futrell, J. H., Reactions of $C_3H_7^+$ with C_3 and C_4 paraffins, *J. Chem. Phys.*, 45, 560, 1966.

83. Lias, S. G. and Ausloos, P., Structure and reactivity of $C_4H_8^+$ ions formed in the radiolysis of cycloalkanes in the gas phase, *J. Am. Chem. Soc.*, 92, 1840, 1970.

84. Sieck, L. W. and Searles, S. K., High pressure photoionization mass spectrometry. II. A study of thermal $H^-(H°)$ and $H_2^-(H_2°)$ transfer reactions occurring in alkane-olefin mixtures, *J. Am. Chem. Soc.*, 92, 2937, 1970.

85. McMahon, T. B., Blint, R. J., Ridge, D. P., and Beauchamp, J. L., Determination of carbonium ion stabilities by ion cyclotron resonance spectroscopy, *J. Am. Chem. Soc.*, 94, 8934, 1972.

86. Dawson, J. H. J., Henderson, W. G., O'Malley, R. M., and Jennings, K. R., Halide ion transfer reactions in the gas phase chemistry of haloalkanes, *Int. J. Mass Spectrom. Ion Phys.*, 11, 61, 1973.

87. Blint, R. J., McMahon, T. B., and Beauchamp, J. L., Gas-phase ion chemistry of fluoromethanes by ion cyclotron resonance spectroscopy. New techniques for the determination of carbonium ion stabilities, *J. Am. Chem. Soc.*, 96, 1269, 1974.

88. Wieting, R. D., Staley, R. H., and Beauchamp, J. L., Relative stabilities of carbonium ions in the gas phase and solution. A comparison of cyclic and acyclic alkyl carbonium ions, acyl cations, and cyclic halonium ions, *J. Am. Chem. Soc.*, 96, 7552, 1974.

89. Sieck, L. W., Lias, S. G., Hellner, L., and Ausloos, P., Photoionization of C_4H_8 isomers. Unimolecular and bimolecular reactions of the $C_4H_8^+$ ion, *J. Res. Natl. Bur. Stand.*, 76A, 115, 1972.

90. Abramson, F. B. and Futrell, J. H., Ionic reactions in unsaturated compounds. III. Propylene and the isomeric butenes, *J. Phys. Chem.*, 72, 1994, 1968.

91. Myher, J. J. and Harrison, A. G., Ion-molecule reactions in propyne and allene, *J. Phys. Chem.*, 72, 1905, 1968.

92. Bowers, M. T., Aue, D. H., and Elleman, D. D., Mechanisms of ion-molecule reactions in propene and cyclopropane, *J. Am. Chem. Soc.*, 94, 4255, 1972.

93. Chin Lin, M. S. and Harrison, A. G., Ion-molecule reactions in isobutene, *Int. J. Mass Spectrom. Ion Phys.*, 17, 97, 1975.

94. Herod, A. A. and Harrison, A. G., Effect of kinetic energy on ionic reactions in propylene and cyclopropane, *J. Phys. Chem.*, 73, 3189, 1969.

95. Hughes, B. M. and Tiernan, T. O., Ionic reactions in gaseous cyclobutane, *J. Chem. Phys.*, 51, 4373, 1969.

96. Sieck, L. W., Searles, S. K., and Ausloos, P., High pressure photoionization mass spectrometry. I. Unimolecular and bimolecular reactions of $C_4H_8^+$ from cyclobutane, *J. Am. Chem. Soc.*, 91, 7627, 1969.

97. Henis, J. M. S., Ion-molecule reactions in olefins, *J. Chem. Phys.*, 52, 282, 1970.

98. Gross, M. L. and Norbeck, P., Effects of vibrational energy on the rates of ion-molecule reactions, *J. Chem. Phys.*, 54, 3651, 1971.

99. Sieck, L. W. and Ausloos, P., Photoionization of propylene, cyclopropane, and ethylene. The effect of internal energy on the bimolecular reactions of $C_2H_4^+$ and $C_3H_6^+$, *J. Res. Natl. Bur. Stand.*, 76A, 253, 1972.

100. Bowers, M. T., Elleman, D. D., O'Malley, R. M., and Jennings, K. R., Analysis of ion-molecule reactions in allene and propyne by ion cyclotron resonance, *J. Phys. Chem.*, 74, 2583, 1970.

101. Ferrer-Correia, A. J. and Jennings, K. R., ICR mass spectra of fluoroalkenes. IV. Ion-molecule reactions in mixtures of ethylene and the fluoroethylenes, *Int. J. Mass Spectrom. Ion Phys.*, 11, 111, 1973.

102. Ferrer-Correia, A. J., Jennings, K. R., and Sen Sharma, D. K., The use of ion-molecule reactions in the mass spectrometric location of double bonds, *Org. Mass Spectrom.*, 11, 867, 1976.

103. Chai, R. and Harrison, A. G., Location of double bonds by chemical ionization mass spectrometry, *Anal. Chem.*, 53, 34, 1981.

104. Herod, A. A. and Harrison, A. G., Bimolecular reactions of ions trapped in an electron space charge, *Int. J. Mass Spectrom. Ion Phys.*, 4, 415, 1970.

105. Kebarle, P., Higher order reactions, ion clusters and ion solvation, in *Ion-Molecule Reactions*, Franklin, J. L., Ed., Plenum Press, New York, 1972.

106. Good, A., Durden, D. A., and Kebarle, P., Ion-molecule reactions in pure nitrogen and nitrogen containing traces of water at total pressures of 0.5-4 torr. Kinetics of clustering reactions forming $H^+(H_2O)_n$, *J. Chem. Phys.*, 52, 212, 1970.

107. Paulson, J. F., Mosher, R. L., and Dale, F., Fast ion-molecule reactions in CO_2, *J. Chem. Phys.*, 44, 3025, 1966.

108. Bohme, D. K. and Fehsenfield, F. C., Thermal reactions of O_2^+ and O^- ions in gaseous ammonia, *Can. J. Chem.*, 47, 2715, 1969.

109. Ferguson, E. E., Fehsenfeld, F. C., and Albritton, D. L., Ion chemistry of the Earth's atmosphere, in *Gas Phase Ion Chemistry*, Bowers, M. T., Ed., Academic Press, New York, 1979.

110. Melton, C. E., *Principles of Mass Spectrometry and Negative Ions*, Marcel Dekker, New York, 1970.

111. Dillard, J. G., Negative ion mass spectrometry, *Chem. Rev.*, 73, 589, 1973.

112. Christophoru, L. G., The lifetimes of metastable negative ions, *Adv. Electron. Electron Phys.*, 46, 55, 1978.

113. Ho, A. G., Bowie, J. H., and Fry, A., Electron impact studies. LVII. Negative ion mass spectrometry of functional groups: simple esters, *J. Chem. Soc. B*, 530, 1971.

114. Bowie, J. H., Electron impact studies. LXXXII. Negative ion mass spectrometry of functional groups. Collision induced spectra of the carboxyl group, *Org. Mass Spectrom.*, 9, 304, 1974.

115. Bowie, J. H. and Janposri, S., Electron impact studies. LX. Negative ion mass spectrometry of functional groups. Skeletal rearrangements in aryl-nitro compounds, *Org. Mass Spectrom.*, 5, 945, 1971.

116. Bowie, J. H. and Janposri, S., Electron impact studies. CI. Negative ion mass spectra of the carbonyl group, the aryl-CH_2-CO- and aryl-$(CH_2)_2$-CO-systems, *Org. Mass Spectrom.*, 10, 1117, 1975.

117. Christophoru, L. G. and Stockdale, J. A. D., Dissociative electron attachment to molecules, *J. Chem. Phys.*, 48, 1956, 1968.

118. Hunt, D. F., Stafford, G. C., Crow, F. W., and Russell, J. W., Pulsed positive negative ion chemical ionization mass spectrometry, *Anal. Chem.*, 48, 2098, 1976.

119. Hunt, D. F. and Crow, F. W., Electron capture negative ion chemical ionization mass spectrometry, *Anal. Chem.*, 50, 1781, 1978.

120. Fehsenfeld, F. C., Associative detachment, in *Interactions Between Ions and Molecules*, Ausloos, P., Ed., Plenum Press, New York, 1975.
121. Fehsenfeld, F. C., Ferguson, E. E., and Schmeltekopf, A. L., Thermal energy associative detachment reactions of negative ions, *J. Chem. Phys.*, 45, 1844, 1966.
122. Ferguson, E. E., Fehsenfeld, F. C., and Schmeltekopf, A. L., Ion molecule reaction rates measured in a discharge afterglow, *Adv. Chem. Ser.*, 80, 83, 1969.
123. Lieder, C. A. and Brauman, J. I., A technique for detection of neutral products in gas-phase ion-molecule reactions, *Int. J. Mass Spectrom. Ion Phys.*, 16, 307. 1975.
124. Lieder, C. A. and Brauman, J. I., Detection of neutral products in gas-phase ion-molecule reactions, *J. Am. Chem. Soc.*, 96, 4028, 1974.
125. Olmstead, W. N. and Brauman, J. I., Gas-phase nucleophilic displacement reactions, *J. Am. Chem. Soc.*, 99, 4219, 1977.
126. Bohme, D. K. and Young, L. B., Kinetic studies of the reactions of oxide, hydroxide, alkoxide, phenyl, and benzylic anions with methyl chloride in the gas phase at 22.5°, *J. Am. Chem. Soc.*, 92, 7354, 1970.
127. Tanaka, K., Mackay, G. I., Payzant, J. D., and Bohme, D. K., Gas-phase reactions of anions with halogenated methanes at 297 ± 2K, *Can. J. Chem.*, 54, 1643, 1976.
128. Pellerite, M. J. and Brauman, J. I., Intrinsic barriers in nucleophilic displacements, *J. Am. Chem. Soc.*, 102, 5993, 1980.
129. Brauman, J. I., Factors influencing thermal ion-molecule rate constants, in *Kinetics of Ion-Molecule Reactions*, Ausloos, P., Ed., Plenum Press, New York, 1979.
130. Bartmess, J. E. and McIver, R. T., The gas-phase acidity scale, in *Gas Phase Ion Chemistry*, Bowers, M. T., Ed., Academic Press, New York, 1979.
131. Bohme, D. K., Ion chemistry: a new perspective, *Trans. R. Soc. Can.*, in press.
132. Farneth, W. E. and Brauman, J. I., Dynamics of proton transfer involving delocalized negative ions in the gas phase, *J. Am. Chem. Soc.*, 98, 7891, 1976.
133. Futrell, J. H. and Tiernan, T. O., Tandem mass spectrometric studies of ion-molecule reactions, in *Ion-Molecule Reactions*, Franklin, J. L., Ed., Plenum Press, New York, 1972.
134. Goode, G. C. and Jennings, K. R., Reactions of O⁻ ions with some unsaturated hydrocarbons, *Adv. Mass Spectrom.*, 6, 797, 1974.
135. Dawson, J. H. J. and Jennings, K. R., Production of gas-phase radical anions by reaction of O⁻ with organic substrates, *J. Chem. Soc. Faraday Trans. II*, 72, 700, 1976.
136. Harrison, A. G. and Jennings, K. R., Reactions of O⁻ with carbonyl compounds, *J. Chem. Soc. Faraday Trans. II*, 72, 1601, 1976.
137. Rutherford, J. A. and Turner, B. R., The production of NO₂⁻ by electron transfer from O⁻, O₂⁻, O₃⁻ and OH⁻ to NO₂, *J. Geophys. Res.*, 72, 3795, 1967.
138. Fehsenfeld, F. C., Schmeltekopf, A. L., Schiff, H. I., and Ferguson, E. E., Laboratory measurements of negative ion reactions of atmospheric interest, *Planet. Space Sci.*, 15, 373, 1967.
139. Tiernan, T. O., Reactions of negative ions, in *Interactions Between Ions and Molecules*, Ausloos, P., Ed., Plenum Press, New York, 1975.
140. Beaty, E. C., Branscomb, L. M., and Patterson, P. L., Mobilities of oxygen negative ions in oxygen, *Bull. Am. Phys. Soc.*, 9, 535, 1964.
141. Moruzzi, J. L. and Phelps, A. V., Survey of negative ion-molecule reactions in O₂, CO₂, H₂O, CO and mixtures of these gases at high pressures, *J. Chem. Phys.*, 45, 4617, 1966.
142. Fehsenfeld, F. C. and Ferguson, E. E., Laboratory studies of negative ion reactions with atmospheric trace constituents, *J. Chem. Phys.*, 61, 3181, 1974.
143. Field, F. H., Chemical ionization mass spectrometry, in *Mass Spectrometry*, MTP Rev. Sci. Phys. Chem., Vol. 5, Ser. 1, Maccoll, A., Ed., Butterworths, London, 1972.
144. Spinks, J. W. T. and Woods, R. J., *An Introduction to Radiation Chemistry*, John Wiley & Sons, New York, 1964.
145. Harrison, A. G., Jones, E. G., Gupta, S. K., and Nagy, G. P., Total cross sections for ionization by electron impact, *Can. J. Chem.*, 44, 1967, 1966.
146. Brønsted, J. N., Einize bemerkungen über den begriff der sauren und basen, *Recl. Trav. Chim. Pay-Bas.*, 42, 718, 1923.
147. Beauchamp, J. L. and Buttrill, S. E., Proton affinities of H₂S and H₂O, *J. Chem. Phys.*, 48, 1783, 1968.
148. Long, J. and Munson, B., Proton affinities of some oxygenated compounds, *J. Am. Chem. Soc.*, 95, 2427, 1973.
149. Aue, D. H., Webb, H. M., and Bowers, M. T., Quantitative proton affinities, ionization potentials, and hydrogen affinities of alkyl amines, *J. Am. Chem. Soc.*, 98, 311, 1976.
150. Briggs, J. P., Yamdagni, R., and Kebarle, P., Intrinsic basicities of ammonia, methyl amines, anilines, and pyridine from gas-phase proton-exchange equilibria, *J. Am. Chem. Soc.*, 94, 5128, 1972.

151. Yamdagni, R. and Kebarle, P., Gas-phase basicities of amines. Hydrogen bonding in proton-bound amine dimers and proton-induced cyclization of α,ω-diamines, *J. Am. Chem. Soc.*, 95, 3504, 1973.
152. Schiff, H. I. and Bohme, D. K., Flowing afterglow studies at York University, *Int. J. Mass Spectrom. Ion Phys.*, 16, 167, 1975.
153. Lau, Y. K. and Kebarle, P., Substituent effects on the intrinsic basicity of benzene: proton affinities of substituted benzenes, *J. Am. Chem. Soc.*, 98, 7452, 1976.
154. Wren, A., Gilbert, P., and Bowers, M. T., New design of an ion cyclotron resonance cell capable of temperature variation over the range $80 \leqslant T \leqslant 450$ K, *Rev. Sci. Instrum.*, 49, 531, 1978.
155. Taft, R. W., Gas-phase proton-transfer equilibria, in *Proton Transfer Reactions,* Caldin, E. F. and Gold, V., Eds., Chapman and Hall, London, 1975.
156. Aue, D. H. and Bowers, M. T., Stabilities of positive ions from equilibrium gas-phase basicity measurements, in *Gas Phase Ion Chemistry,* Bowers, M. T., Ed., Academic Press, New York, 1979.
157. Hehre, W. J., McIver, R. T., Pople, J. A., and Schleyer, P. v. R., Alkyl substituent effects on the stability of protonated benzenes, *J. Am. Chem. Soc.*, 96, 7162, 1974.
158. Carroll, T. X., Smith, S. R., and Thomas, T. D., Correlations between proton affinity and core electron ionization potentials for double-bonded oxygen. Site of protonation in esters, *J. Am. Chem. Soc.*, 97, 659, 1975.
159. Mills, B. E., Martin, R. L., and Shirley, D. A., Further studies of core binding energy-proton affinity correlations in molecules, *J. Am. Chem. Soc.*, 98, 2380, 1976.
160. Benoit, F. M. and Harrison, A. G., Predictive value of proton affinity-ionization energy correlations involving oxygenated molecules, *J. Am. Chem. Soc.*, 99, 3980, 1977.
161. Brown, R. S. and Tse, A., Determination of circumstances under which the correlation of core binding energy and gas-phase basicity or proton affinity breaks down, *J. Am. Chem. Soc.*, 102, 5222, 1980.
162. Liauw, W. G. and Harrison, A. G., Site of protonation in the reaction of gaseous Brønsted acids with halobenzene derivatives, *Org. Mass Spectrom.*, 16, 388, 1981.
163. Tsang, C. W. and Harrison, A. G., The chemical ionization of amino acids, *J. Am. Chem. Soc.*, 98, 1301, 1976.
164. Franklin, J. L., Dillard, J. G., Rosenstock, H. M., Herron, J. T., Draxl, K., and Field, F. H., Ionization Potentials, Appearance Potentials, and Heats of Formation of Gaseous Positive Ions, NSRDS-NBS 28, U.S. National Bureau of Standards, Washington, D.C., 1969.
165. Solomon, J. J. and Field, F. H., Stability of some C_7 tertiary alkyl carbonium ions, *J. Am. Chem. Soc.*, 98, 1025, 1976.
166. Solomon, J. J. and Field, F. H., Reversible reactions of gaseous ions. X. The intrinsic stability of the norbornyl cation, *J. Am. Chem. Soc.*, 98, 1567, 1976.
167. Cox, J. D. and Pilcher, G., *Thermochemistry of Organic and Organometallic Compounds,* Academic Press, New York, 1970.
168. Stull, D. F. and Prophet, H., JANAF Thermochemical Tables, National Standards Ref. Data Ser., U.S. National Bureau of Standards, Washington, D.C., 1971.
169. Smyth, K. C. and Brauman, J. I., Photodetachment of electrons from phosphide ion; the electron affinity of PH_2, *J. Chem. Phys.*, 56, 1132, 1972.
170. Lineberger, W. C., Laser photodetachment electron spectroscopy, in *Laser Spectroscopy,* Brewer, R. and Mooradian, A., Eds., Plenum Press, New York, 1974.
171. Lifshitz, C., Tiernan, T. O., and Hughes, B. M., Electron affinities from endothermic negative-ion charge-transfer reactions, NO_2 and SF_6, *Chem. Phys. Lett.*, 7, 469, 1970.
172. Janousek, B. K. and Brauman, J. I., Electron affinities, in *Gas Phase Ion Chemistry,* Bowers, M. T., Ed., Academic Press, New York, 1979.
173. Hotop, H. and Lineberger, W. C., Binding energies in atomic negative ions, *J. Phys. Chem. Ref. Data*, 4, 539, 1975.
174. Celotta, R. J., Bennett, R. A., and Hall, J. L., Laser photodetachment determination of the electron affinities of OH, NH_2, NH, SO_2, and S_2, *J. Chem. Phys.*, 60, 1740, 1974.
175. Hughes, B. M., Lifshitz, C., and Tiernan, T. O., Electron affinities from endothermic negative-ion change-transfer reactions. III. NO, NO_2, SO_2, CS_2, Cl_2, Br_2, I_2, and C_2H, *J. Chem. Phys.*, 59, 3162, 1973.
176. Chupka, W. A., Berkowitz, J., and Gutman, D., Electron affinities of halogen diatomic molecules as determined by endoergic charge transfer, *J. Chem. Phys.*, 55, 2724, 1971.
177. Engelking, P. C., Ellison, G. B., and Lineberger, W. C., Laser photodetachment electron spectroscopy of methoxide, deuteromethoxide, and thiomethoxide: electron affinities and vibrational structure of CH_3O, CD_3O, and CH_3S, *J. Chem. Phys.*, 69, 1826, 1978.
178. Zimmerman, A. H. and Brauman, J. I., Electron photodetachment from negative ions of C_{2v} symmetry. Electron affinities of allyl and cyanomethyl radicals, *J. Am. Chem. Soc.*, 99, 3565, 1977.

179. Engelking, P. C. and Lineberger, W. C., Laser photodetachment spectrometry of $C_5H_5^-$: a determination of the electron affinity and Jahn-Teller coupling in cyclopentadienyl, *J. Chem. Phys.*, 67, 1412, 1977.

180. Zimmerman, A. H., Reed, K. J., and Brauman, J. I., Photodetachment of electrons from enolate anions. Gas phase electron affinities of enolate radicals, *J. Am. Chem. Soc.*, 99, 7203, 1977.

181. Christophoru, L. G. and Goans, R. E., Low energy (<1 eV) electron attachment to molecules in very high pressure gases: C_6H_6, *J. Chem. Phys.*, 60, 4244, 1974.

182. Lifshiftz, C., Tiernan, T. O., and Hughes, B. M., Electron affinities from endothermic negative ion charge-transfer reactions. IV. SF_6, selected fluorocarbons, and other polyatomic molecules, *J. Chem. Phys.*, 59, 3182, 1973.

183. Wentworth, W. E., Chen, E., and Lovelock, J. E., The pulse sampling technique for the study of electron-attachment phenomena, *J. Phys. Chem.*, 70, 445, 1966.

184. Chaudhuri, J., Jagur-Grodzinski, J., and Szwarc, M., Electron affinities of aromatic hydrocarbons in the gas phase and in solution, *J. Phys. Chem.*, 71, 3063, 1967.

185. Rosenstock, H. M., Draxl, K., Steiner, B. W., and Herron, J. T., Energetics of gaseous ions, *J. Phys. Chem. Ref. Data*, 6(1), 1977.

186. McMahon, T. B. and Kebarle, P., Intrinsic acidities of substituted phenols and benzoic acids determined by gas-phase proton-transfer equilibria, *J. Am. Chem. Soc.*, 99, 2222, 1977.

187. Cumming, J. B. and Kebarle, P., Summary of gas phase acidity measurements involving acids AH. Entropy changes in proton transfer reactions involving negative ions. Bond dissociation energies D(A-H) and electron affinities EA(A), *Can. J. Chem.*, 56, 1, 1978.

188. Bartmess, J. E., Scott, J. A., and McIver, R. T., Scale of acidities in the gas phase from methanol to phenol, *J. Am. Chem. Soc.*, 101, 6046, 1979.

189. Mackay, G. I., Hemsworth, R. S., and Bohme, D. K., Absolute gas phase acidities of CH_3NH_2, $C_2H_5NH_2$, $(CH_3)_2NH$, and $(CH_3)_3N$, *Can. J. Chem.*, 54, 1624, 1976.

190. Bartmess, J. E., Scott, J. A., and McIver, R. T., Substituent and solvation effects in gas-phase acidities, *J. Am. Chem. Soc.*, 101, 6056, 1979.

Chapter 3

INSTRUMENTATION FOR CHEMICAL IONIZATION

I. INTRODUCTION

In chemical ionization mass spectrometry, ionization of the sample of interest is accomplished by interaction of the sample molecules with gaseous reagent ions, viz.,

$$R^{\pm} + S \rightarrow S^{\pm} + R \tag{1}$$

The yield of sample ions $S^{\pm}$ depends on the magnitude of the ion/molecule reaction rate constant, on the concentration (or partial pressures) of the reacting species $R^{\pm}$ and S, and on the time available for reaction between the two. As discussed in the previous chapter the reaction rate constant frequently is close to the limiting value determined by the ion/molecule collision rate as established by ion/dipole and ion/induced dipole interactions. The partial pressures of reacting species required and the reaction time available are interrelated quantities which depend on the type of instrument used. Instrumental features peculiar to chemical ionization mass spectrometry are discussed in this chapter; no attempt is made to cover in detail the instrumentation of mass spectrometry in general, which has been adequately reviewed elsewhere.[1-4]

II. MEDIUM PRESSURE MASS SPECTROMETRY

The majority of chemical ionization studies have been carried out using ion sources only slightly modified from those used in electron impact studies; indeed most are capable of operating in either the electron impact or chemical ionization mode. In such sources diffusional loss of ions to the walls and removal of ions from the source by the necessary extraction fields limits the source residence time to no more than approximately 10 μsec. To achieve adequate sensitivity in the chemical ionization mode it is necessary to operate with reagent gas pressures in the range 0.5 to 2.0 torr and to use sample pressures of the order of 10^{-3} to 10^{-4} torr in the ion source. At the same time the mass analysis region must be maintained in the region of 10^{-6} torr to prevent excessive ion scattering and the pressure in the ion source housing must be maintained at $\leqslant 10^{-4}$ torr to prevent damage to the electron-emitting filament and to prevent ion/molecule collisions between the source ion exit slit and the point of entry into the mass analysis region. In addition, for magnetic deflection instruments, the pressure in the source housing must be kept at this level to prevent discharge to ground of the high accelerating potential that normally is used.

To achieve these conditions the ion source must be made as gas-tight as possible, the pumping speed at the source housing must be as high as practicable, and the analyzer region must be differentially pumped. The ion source is carefully constructed to ensure a gas-tight assembly and the aperture for entry of the ionizing electron beam is reduced relative to that used in electron impact studies, typically to 0.1 to 0.2 mm². The ion exit slit is reduced both in width and length, typical widths being of the order of 5×10^{-2} mm and total area 0.1 to 0.2 mm²; a somewhat larger slit area can be accommodated if a thicker end plate is used to limit the gas conductance of the slit. Apart from gas-tight connections for reagent gas and sample introduction these normally are the only apertures in the ion source. With such a source and high-speed diffusion or turbomolecular pumps, source pressures of ∿1 torr can be maintained with source housing pressures <10^{-4} torr and analyzer pressures of ∿10^{-6} torr. Although most manufacturers now offer a combined chemical ionization/electron impact source as a standard feature and, therefore, have incorporated at least adequate pumps on their instruments, a number of reports have appeared giving details of the modification of existing electron impact mass spectrometers for operation in the chemical ionization mode.[5-10] Usually included in the modifications has been a redesign of the source housing and pumping leads to provide adequate source region pumping speeds; the simple addition of a high-speed diffusion pump is not adequate if the pumping line leading to the source is of small diameter, as is frequently the case in older instruments.

With a source pressure of 0.5 to 1.0 torr, electrons with 70 eV energy will have a very short range and it is usual to increase the electron energy to about 400 eV in order that the electrons will penetrate further into the source. However, even under these conditions the electron beam will be completely attenuated before it reaches the position where the electron trap normally is installed in an electron impact source. Consequently, the filament current cannot be regulated to give a constant trap current as is done in electron impact operation but rather the filament current is regulated to give a constant electron current to the source block. It is hoped that if a constant electron flux impinges on the source block a constant flux will pass through the electron entrance aperture and enter the ionization chamber.

Electron impact ionization of the reagent gas normally is employed, and with most commonly used reagent gases the increased pressure does not lead to an unacceptably short filament lifetime. However, if oxidizing reagents such as nitric oxide or oxygen are used as reagents, filament lifetimes are severely reduced. For these cases Hunt and co-workers[11] have designed a Townsend discharge ion source, illustrated schematically in Figure 1, to replace the electron impact source. The discharge occurs between two screen electrodes 1/8 in. apart, one in contact with the source block and one in contact with the stainless-steel gas inlet line. In the normal mode of operation the metal screen at source potential serves as the anode and discharge occurs when the cathode (and gas inlet line) are 600 to 1300 V negative with respect to the anode. Under these conditions electrons produced in the discharge are accelerated through the anode screen (85% transmission) into the ion source where the reagent gas is ionized by electron impact. Both positive and negative ion chemical ionization systems have been investigated and give reagent ion spectra very similar to those produced by electron impact. Alternatively, if the source screen is made the cathode and the anode screen and inlet line are maintained 600 to 1300 volts positive with respect to the source, positive ions produced in the discharge are accelerated through the screen into the ion source. Reagent gas spectra produced in this manner showed lower intensities for cluster ions, such as $NH_4^+(NH_3)_2$ in ammonia, presumably because of the higher kinetic energy of the reactant ions.

A problem associated with the operation of magnetic sector instruments under chemical ionization conditions has been that of electrical discharges through the gas at 0.5 to 1.0 torr between the ion source which is at a high positive or negative potential with respect to ground and a ground located in the sample inlet lines. This discharge may occur in the line admitting the reagent gas, in the line admitting volatile samples, or in the line leading to a pressure measuring device. The discharge occurs when the electric field causes a breakdown of the sample gas resulting in a flow of electrons toward the anode (the ion source is positive ion CI) and of positive ions toward the cathode (ground in positive ion CI). The breakdown voltage is a function of the electric field strength and the gas density and varies considerably with the nature of the gas; breakdown is particularly prevalent, however, at pressures used in chemical ionization. The discharge is quenched at higher pressures (above *circa* 50 torr) and this fact has been used in several inlet systems where a large pressure drop is created through a noncon-

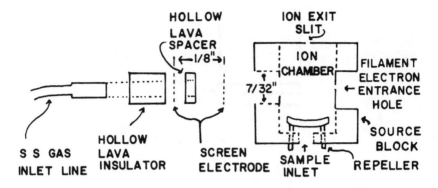

FIGURE 1. Townsend discharge ion source. (From Hunt, D. F., McEwen, C. N., and Harvey, T. M., *Anal. Chem.*, 47, 1730, 1975. With permission.)

ducting capillary leak. The disadvantage of such an approach is that large sample backing pressures are required, creating a problem when small sample sizes or compounds of low vapor pressure are used; in addition it is impossible to measure the source pressure directly through such a lead. Futrell and Wojcik[6] have solved the discharge problem by using a chain of vacuum resistors between the ion source and ground, with stainless-steel turnings at each potential, to create a uniform potential gradient between the source and ground. More recently, Illes et al.[12] have reported that packing a glass inlet line with approximately 4 in. of glass wool prevents discharges through the line. Presumably the glass wool serves to reduce the positive ion velocities below that required for secondary electron emission and/or provides a very large surface area to act as a third body in ion/electron recombinations. In an alternative approach[13] the entire inlet system and pressure measuring device have been floated at source potential and powered via a main isolation transformer while being located behind a protective screen. An interesting point is that Pyrex® glass becomes a good conductor for electricity at 250 to 300°C and it may be necessary to insert a short length of tubing made from quartz or other insulating glass in the line. These discharge problems are avoided when quadrupole mass spectrometers are used since much lower potentials are applied to the ion source and focusing electrodes.

When a magnetic deflection instrument is operated under negative chemical ionization conditions negative ions can be readily detected using a conventional electron multiplier if the potential of the ion source is appreciably more negative with respect to earth than the potential of the first dynode of the electron multiplier. For example, a negative ion accelerated from an ion source at −8 kV will still have 5 keV of kinetic energy when it collides with the first dynode of an electron multiplier held at −3 kV. However, if the two voltages are similar or if a quadrupole mass filter is used, where the ions usually have <100 eV of kinetic energy no ion signal will be detected unless the detection system is modified. If the first dynode of the multiplier is held at ground potential or at a positive potential and the last dynode typically is held at +2 to +6 kV, secondary electrons will be produced. However, the output signal must be conducted by a suitable high-voltage, floating, coaxial signal feed-through to a preamplifier and appropriate circuiting to reference the signal to ground potential.[14,15] While this is possible the approach has the associated problem that stray electrons in the vacuum system are detected efficiently, causing an increase in the background signal by a factor of approximately 30.[15] An alternative approach, proposed by Stafford and co-workers,[15,16] is to convert the negative ions to positive ions by impact on a conversion dynode biased at a suitably high positive potential. The positive ions so produced are then accelerated to the first, negatively biased, dynode of the multiplier and are detected in the usual way; the device is illustrated in Figure 2. It is believed that the

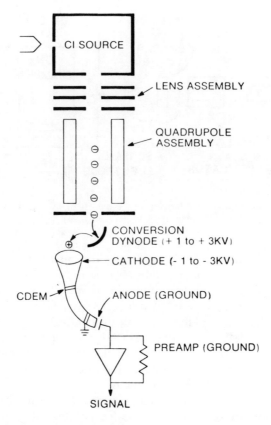

FIGURE 2. Negative-ion continuous dynode electron multiplier, incorporating conversion dynode. (From Stafford, G. C., *Environ. Health Perspect.*, 36, 85, 1980. With permission.)

major mechanisms responsible for the negative-to-positive ion conversion are sputtering, fragmentation, and charge stripping. Normally the gain of an electron multiplier on impact of positive ions decreases with increasing ion mass. By contrast, the gain for negative ions using the conversion dynode multiplier increases with increasing ion mass.[15,16]

Modification of an electron impact ion source for chemical ionization operation leaves a source with which it is possible to obtain electron impact mass spectra by stopping the flow of reagent gas and, in some cases, adjusting the ion source focusing potentials. In the author's experience, the more highly gas-tight ion source leads to higher ion source pressures and consequent ion/molecule reactions involving sample ions. These frequently result in formation of $[M + 1]^+$ ions which can achieve an intensity considerably above the level expected from isotope peaks. Thus, while combined EI/CI sources give electron impact spectra which are satisfactory for identification purposes, better-quality spectra are obtained using an ion source which has been specifically designed for electron impact studies. Similarly a source designed specifically for chemical ionization operates more efficiently. However, the convenience of being able to obtain both types of spectra without changing sources cannot be denied. Along these lines several manufacturers now supply ion sources in which either the ion exit slit or the complete source volume are changed without breaking the vacuum when changing from one ionization condition to the other. In an alternative approach, Arsenault et al.[17] have described a dual source which permits both electron impact and chemical ionization spectra to be recorded on the same sample.

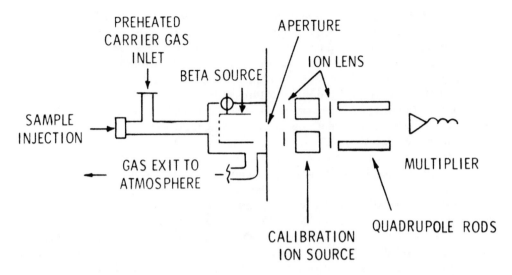

FIGURE 3. Atmospheric pressure ionization mass spectrometer. (From Horning, E. C., Horning, M. G., Carroll, D. I., Dzidic, I., and Stillwell, R. N., *Anal. Chem.*, 45, 936, 1973. With permission.)

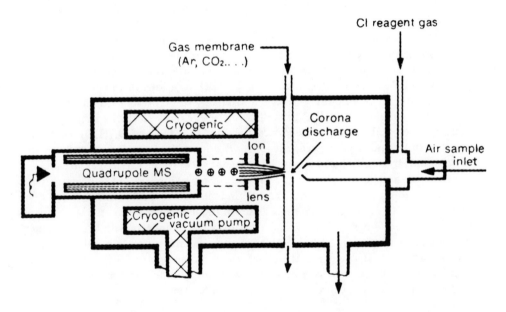

FIGURE 4. TAGA® atmospheric pressure ionization mass spectrometer. (From Lane, D. A., Thomson, B. A., Lovett, A. M., and Reid, N. M., *Adv. Mass Spectrom.*, 8, 1480, 1980. With permission.)

III. ATMOSPHERIC PRESSURE IONIZATION (API) MASS SPECTROMETRY

In atmospheric pressure ionization (API) mass spectrometry, positive and negative ions are generated in a flowing carrier gas at atmospheric pressure either by β-particles from a ^{63}Ni source[18,19] or by a corona discharge.[20,21] The ions present in the flowing gas subsequently are sampled through a small aperture and focused by a suitable ion lens into a quadrupole mass filter for analysis. Schematic diagrams of typical API systems are shown in Figures 3 and 4. Samples are introduced into the flowing gas stream by injection in a solvent, by direct insertion on a probe, or as an effluent from a gas or liquid chromatograph; if ambient air is used as the carrier gas trace impurities in the air can be detected directly. A considerable amount of gas passes through the aperture and to maintain the mass spectrometer at a suitable pressure ($<10^{-5}$ torr) requires efficient pumping. A unique feature of the design shown in Figure 4 is the use of cryogenic pumping to give effective pumping speeds of 2×10^4 ℓsec^{-1} at an operating pressure of 10^{-5} torr. A second unique feature of this instrument is the use of an inert gas membrane flowing past the aperture through which the ions can pass, while neutral species, including particulate matter which might clog the orifice, are rejected.

Theoretical considerations[22,23] show that the terminal ions formed in the weak plasma produced using a ^{63}Ni source often are dominated by ions characteristic of trace impurities rather than by the major component(s). In effect, concentration and reaction time parameters are such that the ionization ends up residing in the most stable positive or negative ionic species. The higher ionization rates in corona discharge API instruments precludes a similar theoretical treatment although the same general conclusions probably apply. Since many species of interest in atmospheric, biological, or environmental studies have either large electron affinities or large proton affinities API is of interest as an ultrasensitive trace analyzer. The efficiency (ions out per sample molecule in) of such an instrument has been estimated[22,23] to be between 10^{-3} and 10^{-1} and to be independent of engineering details, creating the possibility of achieving detection levels in the part-per-trillion range in suitable cases. Experimental results support this conclusion.

In a rigorously cleaned API instrument using extremely high-purity nitrogen carrier gas the major primary charged species are N_3^+ and N_4^+ as positive ions and thermal energy electrons as negative species. In practical operation trace amounts of water will be present leading to appreciable intensities for $H^+(H_2O)_n$ ions in the positive ion regime. In addition, trace amounts of oxygen are likely to be present leading to formation of O_2^- (and possibly $O_2^-(H_2O)_n$) as well as thermal energy electrons in the negative ion mode.[24] When ambient air is used as carrier gas $H^+(H_2O)_n$ and $O_2^-(H_2O)_n$ constitute the major positive and negative reagent ions. Of course, it is possible to prepare specific reagent ions by addition of appropriate reagent gases to the carrier gas stream.

The major advantage of atmospheric pressure chemical ionization appears to lie in its extremely high ultimate sensitivity, and it finds its major application in trace analysis. A possible difficulty of the method arises in the identification of the components of complex mixtures where the final ionization may reside entirely with the component forming the most stable ionic products and components forming less stable products may not be detected. It is clear that in such cases either separations, such as by gas chromatography, will be necessary or selective chemical ionization reagents will need to be developed to ionize specific components in the mixture.

IV. ION CYCLOTRON RESONANCE SPECTROMETRY

Although the majority of chemical ionization studies have been carried out at pressures of $\sim$1 torr in medium pressure sources (or in API instruments), similar studies, in principle, could be carried out at much lower pressures if the reaction time could be made correspondingly longer. Devices in which long reaction times for ion/molecule reactions can be achieved include the quadrupole ion storage trap,[25] the electron space charge trap,[26] and the ion cyclotron resonance (ICR) spectrometer.[27] Although all three techniques have been extensively used to study ion/molecule reaction kinetics and equilibria, only the ICR approach has received significant study as a possible approach to chemical ionization. In a conventional drift cell in ICR the ion residence time is $\sim$10^{-3} sec while in trapped ion cells ions can be stored for periods up to the order of seconds. Compared to the $\sim$10^{-5} sec reaction time available in high pressure chemical ionization it is apparent that partial pressures can be reduced by a factor of 10^5 to 10^6 and still achieve the same extent of conversion in bimolecular ion/molecule reactions in the ICR experiment.

Early studies of the analytical uses of the ICR method showed[28-30] that the technique suffered from several severe limitations. Mass resolution was limited to about 1 amu at m/z 200 while the mass range was restricted to less than about m/z 280. In addition, the scan rate was very low, taking typically 5 min to scan from m/z 10 to m/z 110. Recent developments employing Fourier transform techniques and a one-region ICR cell have largely removed these limitations and provide a mass range and resolution greater than 1000 with complete mass spectra being obtained in the order of seconds rather than minutes. In the rapid-scan ICR technique developed by Hunter and Mc-Iver[31,32] ions trapped in a single cell are mass analyzed by detecting, with a capacitance bridge detector, the transient response of resonant ions in the cell when an excitation frequency from a frequency synthesizer is canned rapidly (over a period of seconds) across the spectrum under computer control. The true mass spectrum is recovered from the detected signal using Fourier transform computation methods. In the Fourier trans-form-ion cyclotron resonance (FT-ICR) technique, first introduced by Comisarow and Marshall,[33] and recently applied to chemical ionization by Gross and co-workers,[34] the excitation frequency sweep is much more rapid (milliseconds) and the transient reso-nant response of all ions is detected simultaneously and converted to a mass spectrum by Fourier transform methods. Both approaches appear to have overcome the major limitations of the conventional ICR techniques.

Using these techniques both Hunter and McIver[32] and Gross and co-workers[34] have demonstrated the feasibility of obtaining chemical ionization mass spectra with the reagent gas in the low 10^{-6} torr range and the analyte in the 10^{-7} to 10^{-8} torr range. However, at the present stage of development, it appears that the analyte pressure must be $\sim 1\%$ of the reagent gas pressure; as a result significant electron impact ioni-zation of the analyte and/or further reaction of analyte ions does occur. The methods are in the early stages of development and further advances permitting operation at lower analyte pressures undoubtedly will be forthcoming.

V. PULSED POSITIVE ION-NEGATIVE ION CHEMICAL IONIZATION

Studies in the last few years have shown that negative ion chemical ionization mass spectrometry has considerable potential as an analytical method.[35] In many cases it offers increased sensitivity and selectivity and, frequently, it provides information which is complementary to positive ion chemical ionization. There is an obvious ad-vantage in obtaining both the positive ion and the negative ion chemical ionization mass spectra of a sample. The quadrupole mass filter transmits negative ions and pos-itive ions with equal facility and Hunt et al.[36] have described a pulsed mode of opera-tion which permits simultaneous recording of positive and negative chemical ionization mass spectra. Their arrangement is shown in Figure 5. By pulsing both the ion source potential ($\pm$ 4 to 10 V) and the focusing lens potential ($\pm$ 10 to 30 V) at a rate of 10 kHz, packets of positive and negative ions are ejected from the source in rapid succes-sion and enter the quadrupole filter. After mass analysis the positive and negative ions are detected by separate continuous-dynode electron multipliers operating at suitable potentials, with the two signals being recorded simultaneously on separate channels of a light beam oscillograph. An alternative detection system employing two conversion dynodes and a single electron multiplier is shown in Figure 6.[15]

By suitable choice of reagent gas various types of chemical ionization mass spectra may be recorded in the positive and negative ion modes.[36] With nitrogen as the reagent gas, N_2^+, reacting by charge exchange, is the dominant positive ion reacting species,

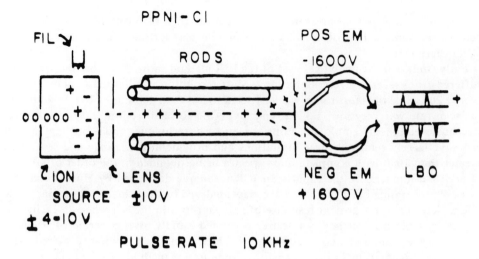

FIGURE 5.   Pulsed positive negative ion chemical ionization (PPNICI) mass spectrometer (FIL — filament, EM — electron multiplier, LBO — light beam oscillograph). (From Hunt, D. F., Stafford, G. C., Crow, F. W., and Russell, J. W., *Anal. Chem.*, 48, 2098, 1976. With permission.)

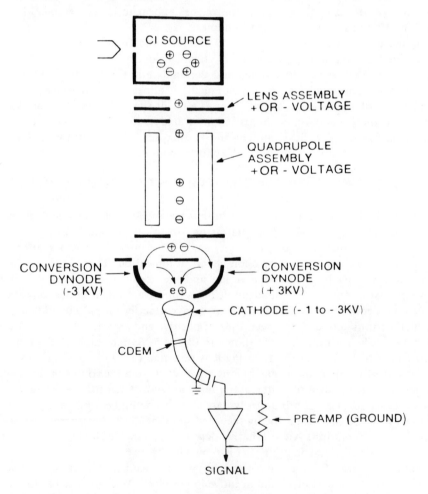

FIGURE 6.   Detector for PPNICI system. (From Stafford, G. C., *Environ. Health Perspect.*, 36, 85, 1980. With permission.)

while quasi-thermal energy electrons provide the ionizing reagent in the negative ion mode. The use of methane as reagent gas gives the Brønsted acid reagents CH_5^+ and $C_2H_5^+$ in the positive ion mode and quasi-thermal energy electrons in the negative ion mode. The addition of CH_3ONO to the methane reagent results in the Brønsted base CH_3O^- in the negative ion mode without changing the positive ion reagents.

This pulsed mode of operation permitting simultaneous recording of positive and negative ion chemical ionization mass spectra is available commercially.

VI. SAMPLE INTRODUCTION IN CHEMICAL IONIZATION MASS SPECTROMETRY

The major requirement in conventional chemical ionization mass spectrometry is that the sample of interest be introduced as a vapor, with a partial pressure $<10^{-3}$ torr, into the ion source containing the reagent gas plasma. Samples can be introduced either from a heated inlet system or from a direct insertion probe; one must ensure that these modes of sample introduction do not provide a route for discharge of the high voltage to ground. When samples are introduced from a gas inlet system the pressure differential between the sample in the inlet system and the reagent gas in the source may not be sufficient to permit significant flow through the flow-regulating leak. In this case it is necessary to add reagent gas to the sample reservoir to create the necessary pressure differential.

The combination of a gas chromatograph coupled to a mass spectrometer has been amply shown to be a powerful analytical tool. The problems of interfacing the gas chromatograph to a mass spectrometer have been discussed thoroughly elsewhere.[37] When using packed columns in electron impact studies one usually employs some type of separator for pressure reduction and sample enrichment. For chemical ionization studies the same arrangement can be used with the chemical ionization reagent gas being added through a separate inlet system. The simplest interface is a direct flow connection, without a separator, with all or part of the gas chromatograph effluent being taken into the mass spectrometer. Modern mass spectrometers designed for chemical ionization have a high pumping speed in the ion source region and can accommodate a large fraction of the gas flow from packed columns; the entire effluent from capillary columns can be handled with ease. In this mode of operation the carrier gas normally serves as the reagent gas for chemical ionization, with additional reagent gas being added if necessary; the latter is necessary when capillary columns are used. Several laboratories[36-41] have reported such directly coupled gas chromatograph-chemical ionization mass spectrometer combinations using either time-of-flight or quadrupole mass spectrometers; these types of instruments avoid the problem of discharge of the high voltage to the grounded gas chromatograph which arises when a magnetic-sector instrument is used. More recently, Hatch and Munson[42] have reported the direct coupling of a gas chromatograph to a magnetic-sector mass spectrometer. Commercial instruments are readily available for gas chromatography/chemical ionization mass spectrometry incorporating both direct-coupling interface and interfaces employing separators.

For mixtures which are not sufficiently volatile for gas chromatography, separation by liquid chromatography is often possible and there is considerable interest in the interfacing of liquid chromatographs to mass spectrometers for the sensitive detection and identification of unknown components. The interfacing of a liquid chromatograph to a mass spectrometer is considerably more difficult than the interfacing of a gas chromatograph and successful operation of liquid chromatograph/mass spectrometer combinations has been achieved only relatively recently.[43] In the approach of Mc-Lafferty and co-workers,[44] a small portion (10 to 100 μl/min) of the liquid chromato-

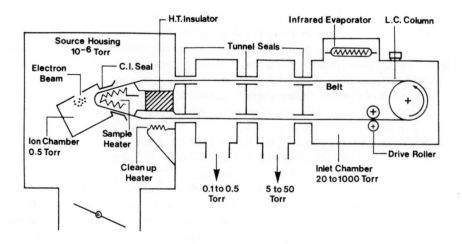

FIGURE 7.   Liquid chromatograph/mass spectrometer moving belt interface. (From Craig, R. D., Bateman, R. H., Green, B. N., and Millington, D. S., *Phil. Trans. R. Soc. London,* A293, 135, 1979. With permission.)

graph effluent is vaporized directly into the ion source of the mass spectrometer where the solvent serves as the reagent gas for chemical ionization of the solute. In general, polar solvents are favored since they more readily yield Brønsted acid-reactant ions which readily can protonate solute molecules. This method suffers from the disadvantages that only a small fraction of the effluent can be accepted into the mass spectrometer, thus limiting sensitivity, and that only chemical ionization mass spectra can be recorded. In addition, the versatility of the chemical ionization system is limited by the fact that the solvent serves as the chemical ionization reagent gas. Some of these problems are absent in the moving belt interface, first developed by McFadden and colleagues.[45] The interface, as adapted for operation with a magnetic sector instrument,[46] is shown schematically in Figure 7. The liquid chromatograph effluent (or part thereof) is carried by a moving belt (2 to 4 cm/sec) through successive vacuum chambers where the volatile solvent is removed and then, by means of a vacuum lock, into the ion source of the mass spectrometer where the sample is evaporated. Following exit from the source the belt is heated more strongly to remove the last of the sample. With this interface it is possible to accommodate up to 0.85 mℓ/min flow from the liquid chromatograph, although this figure is reduced for polar solvents which are less readily vaporized. Using this interface it is possible to obtain either electron impact or chemical ionization mass spectra, with the reagent gas for the latter being added separately. Both approaches are limited in the solute they can handle by the fact that the solute must be vaporized before it is ionized.

The utility of chemical ionization mass spectrometry (and, equally, electron impact and field ionization mass spectrometry) is severely limited for involatile, thermally labile molecules by the requirement that the sample be in the gaseous state prior to ionization and analysis. For such molecules the energy required to overcome the bonding between molecules and/or the bonding of molecules to the sample holder surface may exceed the energy required to break bonds within the sample molecule and, under normal vaporization conditions, extensive sample decomposition may occur. Considerable success in vaporizing intact labile peptides has been achieved by rapid heating (up to 12°C/sec) of samples dispersed on an inert Teflon® surface.[47] The dispersion on the Teflon® surface enhances volatility by decreasing both intermolecular forces and molecule/surface forces. If the heating is sufficiently rapid the kinetic energy of the molecules is increased rapidly and intermolecular forces and molecule/surface forces are overcome before appreciable energy becomes concentrated in intramolecular modes and results in molecular bond rupture. Rather similar results have been obtained by Cotter[48] in studies where rapid heating of the sample dispersed on an inert probe was achieved by using a pulsed laser and ionization was achieved by chemical ionization. In both these approaches the sample duration in the ion source is brief and rapid scanning and data acquisition are essential.

An alternative approach which has been named variously as direct chemical ionization, in-beam ionization, desorption chemical ionization, or surface ionization[49] involves dispersion of the sample on a suitable probe which then is inserted directly into the ion source to a position close to the electron beam. The method has been used in both electron impact and chemical ionization studies. In chemical ionization the method was first employed by Baldwin and McLafferty[50] who used an extended glass probe tip to obtain methane chemical ionization mass spectra of underivatized peptides. Interest in the technique was revived by the work of Hunt et al.[51] who obtained good quality chemical ionization mass spectra of a number of salts and thermally labile organic molecules, including cyclic adensosine monophosphate, using an activated field desorption emitter as a solid probe, with heating of the emitter by passing an electric current through it. Subsequently, several reports of the design of extended probes, frequently heatable and constructed out of inert materials, which extend directly into the reactant ion plasma have been reported;[52-56] in addition most manufacturers of chemical ionization mass spectrometers will supply probes for direct chemical ionization.

The main advantage of the technique is its simplicity. For many compounds, ions indicative of molecular weight are observed where no such ions are seen using conventional solids probes. The mechanism by which ions characteristic of molecular weight are produced using these extended or direct insertion probes is not clear. It is possible that the dispersion of the sample on the inert surface combined with the rapid heating usually employed leads to desorption of neutral molecules and the insertion of the probe directly into the ion plasma results in these desorbed molecules being ionized before they suffer collisions with other surfaces. On the other hand it is possible that ionization of the sample takes place on the surface of the probe tip and that ions are then desorbed from the surface. The available evidence supports the first explanation.[49]

VII. INSTRUMENTATION FOR COLLISION-INDUCED DISSOCIATION (CID) STUDIES

A useful feature of chemical ionization mass spectrometry is that, by the choice of suitable reagent gases, it usually is possible to confine the ionization of the substrate largely, if not completely, to a single ion. This is particularly useful for quantitation of a known compound since maximum sensitivity results. However, when identification of an unknown is required, a single ion spectrum giving molecular weight information is not sufficient; fragment ions providing further structural information usually are required. For pure substrates this can be achieved by using more energetic reagent ions which result in fragment ion formation. However, when one desires to identify the individual components of a complex mixture without prior separation this simple approach fails because the spectrum becomes too complicated to interpret.

The desire to carry out such identifications in complex mixtures without prior separation has led to the development of the technique frequently called mass spectrometry/mass spectrometry (MS/MS).[57-60] In this technique the components of the mixture are ionized, usually by chemical ionization, to produce the simplest possible collection of ions representative of the neutrals present in the mixture. In the situation where one is attempting to identify an individual component, the ion thought to represent that component is selected by using a multistage mass spectrometer and subjected to collision induced dissociation (CID), with the ionic products of the dissociation being identified by mass analysis. Identification is made either by interpretation of the CID spectrum from first principles or by comparison with the CID spectra of pure compounds obtained separately. Each precursor ion in the mixture spectrum can be selected in turn and the component in principle identified by its CID spectrum. In effect, the selection of the reactant ion for CID serves the same purpose as separation of the components by chromatographic techniques while the identification from the CID spectrum is equivalent to identification of the separated components from their EI or CI mass spectra.

These types of study have been carried out extensively using double-focusing mass spectrometers with the configuration B/E (Figure 8a), where the magnetic sector precedes the electrostatic sector. With this instrument, ionization of the sample in the source is followed by mass selection of the ion of interest by the magnetic sector B, with collision-induced dissociation occurring in the collision cell between the two analyzers. The ionic products of dissociation (M_p) will have a kinetic energy

$$T_p = \frac{M_p}{M_i} T_i \qquad (2)$$

where T_i and M_i are the kinetic energy and mass of the precursor ion. The second stage electric sector is a kinetic energy analyzer and by scanning it the masses of the product ions can be established from their kinetic energies relative to the precursor ion kinetic energy. This method represents the so-called MIKES (mass-analyzed ion kinetic energy spectroscopy) technique. Similar results also can be obtained, using either the B/E instrument or the E/B instrument (Figure 8b), by using so-called linked-scan techniques. If the magnetic (B) and electric (E) fields are scanned down at a constant ratio B/E (that necessary to transmit the precursor ion) the ionic products of dissociation of the precursor ion in the collision cell between the source and the first analyzer will be successively detected and can be mass identified from their kinetic energies.[61,62] This linked-scan technique provides a high-resolution spectrum of the product ions in contrast to the normal MIKES spectrum where peaks may overlap due to release of kinetic energy in the fragmentation processes. On the other hand the B/E linked scan is more difficult experimentally and requires a highly accurate measure of the magnetic field strength B with microprocessor control of the scan functions. In an alternative approach the products of fragmentation of a given precursor in the collision cell between the source and first analyzer can be identified by scanning the electric field E and the ion accelerating voltage V keeping the magnetic field B and the ratio E^2/V constant at the value required to transmit the precursor ion.[63,64] Again the mass of the product ion can be determined from its kinetic energy relative to the precursor ion kinetic energy. A disadvantage of this technique is that the ion accelerating voltage is changed causing changing focusing conditions in the ion source.

More recently, triple quadrupole mass spectrometers (Figure 8c) have been developed for such MS/MS studies.[65-67] In these instruments the precursor ion is mass selected by the first quadrupole and is collisionally dissociated in a second quadrupole acting as a collision cell and operating in the rf mode only for ion focusing. The ionic products of dissociation are mass analyzed by the third quadrupole mass analyzer. The major fundamental difference between this technique and those described above lies in the kinetic energies involved in the collisions. In the instruments of Figures 8a or 8b the ions entering the collision cell have energies of the order of keV although only a small fraction of this energy (1 to 25 eV) is converted into internal energy in a "near-

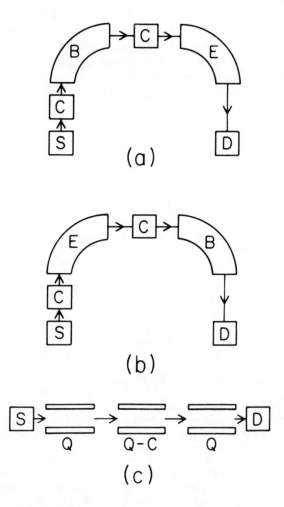

FIGURE 8. Schematic of analyzer arrangements for collision-induced dissociation studies.

miss'' collision with the collision gas. By contrast, in the triple quadrupole experiments the ions entering the collision cell have much lower energies (10 to 200 eV) although it is probable that the energy converted into internal energy in collision is similar.

Two further types of CID studies deserve comment. It frequently is the case that compounds which are closely related structurally will show a common fragment ion in their CID spectra and one may wish to identify all the precursor ions which give this common fragment. This can be accomplished straightforwardly using the triple quadrupole instrument by setting the third quadrupole to transmit the common fragment ion while scanning the first quadrupole to identify the mass of the precursor ions which lead to this common fragment ion on collision-induced dissociation. Similar results can be obtained on double-focusing sector instruments (Figures 8a or 8b) by a linked scan[61] in which the magnetic field and electric field are scanned, while keeping B^2/E constant, to detect the precursors giving the product ion in collision-induced dissociations in the collision cell prior to the first analyzer.

Finally, it sometimes is possible to identify classes of compounds from the mass of the neutral fragment lost on collision-induced dissociation. For example, the carboxylate anions $RCOO^-$ formed in negative ion CI lose CO_2 (mass 44) on CID. Thus, the identification of all precursor ions which lose 44 mass units on collision-induced dissociation serves to identify carboxylic acids. Such a scan is simple in the triple quadrupole instrument where the first and third quadrupoles can be scanned simultaneously but offset by the mass of the neutral lost. A similar scan in double-focusing sector instruments requires a complex linked scan.[68]

Some results from collision-induced dissociation studies of ions formed by chemical ionization are discussed in Chapter 6.

REFERENCES

1. Duckworth, H. E., *Mass Spectroscopy*, Cambridge University Press, Cambridge, 1958.
2. Kiser, R. W., *Introduction to Mass Spectrometry*, Prentice-Hall, Englewood Cliffs, N.J., 1965.
3. Roboz, J., *Introduction to Mass Spectrometry: Techniques and Applications*, Interscience, New York, 1968.
4. Melton, C. E., *Principles of Mass Spectrometry and Negative Ions*, Marcel Dekker, New York, 1970.
5. Michnowicz, J. and Munson, B., Studies in chemical ionization mass spectrometry, *Org. Mass Spectrom.*, 4, 481, 1970.
6. Futrell, J. H. and Wojcik, L. H., Modification of a high resolution mass spectrometer for chemical ionization studies, *Rev. Sci. Instr.*, 42, 244, 1971.
7. Beggs, D., Vestal, M. L., Fales, H. M., and Milne, G. W. A., A chemical ionization mass spectrometer source, *Rev. Sci. Instr.*, 42, 1578, 1971.
8. Hogg, A. M., Modification of an AEI MS-12 mass spectrometer for chemical ionization, *Anal. Chem.*, 44, 227, 1972.
9. Yinon, J. and Boettger, H. G., Modification of an AEI/GEC MS-9 high-resolution mass spectrometer for electron impact/chemical ionization studies, *Chem. Instrum.*, 4, 103, 1972.
10. Garland, W. A., Weinkam, R. J., and Trager, W. F., Relatively simple modification of an AEI MS-902 high-resolution mass spectrometer to chemical ionization studies, *Chem. Instrum.*, 5, 271, 1974.
11. Hunt, D. F., McEwen, C. N., and Harvey, T. M., Positive and negative chemical ionization mass spectrometry using a Townsend discharge ion source, *Anal. Chem.*, 47, 1730, 1975.
12. Illes, A. J., Bowers, M. T., and Meisels, G. G., Sample introduction and pressure measuring system for chemical ionization mass spectrometers, *Anal. Chem.*, 53, 1551, 1981.
13. Jennings, K. R., Chemical ionization mass spectrometry, in *Gas Phase Ion Chemistry*, Bowers, M. T., Ed., Academic Press, New York, 1979.
14. Smit, A. L. C. and Field, F. H., Gaseous anion chemistry. Formation and reactions of OH⁻, Reactions of anions with $N_2O \cdot OH^-$ negative ion chemical ionization, *J. Am. Chem. Soc.*, 99, 6471, 1977.
15. Stafford, G. C., Instrumental aspects of positive and negative ion chemical ionization mass spectrometry, *Environ. Health Perspect.*, 36, 85, 1980.
16. Stafford, G., Reeher, J., Smith, R., and Story, M., A novel negative ion detection system for a quadrupole mass spectrometer, in *Dynamic Mass Spectrometry*, Vol. 5, Price, D. and Todd, J. F. J., Eds., Heyden and Son, London, 1978.
17. Arsenault, G. P., Dolhun, J. J., and Biemann, K., Alternate or simultaneous electron impact-chemical ionization mass spectrometry of gas chromatographic effluent, *Anal. Chem.*, 43, 1720, 1971.
18. Horning, E. C., Horning, M. G., Carroll, D. I., Dzidic, I., and Stillwell, R. N., New picogram detection system based on a mass spectrometer with an external ionization source at atmospheric pressure, *Anal. Chem.*, 45, 936, 1973.
19. Carroll, D. I., Dzidic, I., Stillwell, R. N., Horning, M. G., and Horning, E. C., Subpicogram detection system for gas phase analysis based upon atmospheric pressure ionization (API) mass spectrometry, *Anal. Chem.*, 46, 706, 1974.
20. Dzidic, I., Carroll, D. I., Stillwell, R. N., and Horning, E. C., Comparison of positive ions formed in nickel-63 and corona discharges using nitrogen, argon, isobutane, ammonia, and nitric oxide as reagents in atmospheric pressure ionization mass spectrometry, *Anal. Chem.*, 48, 1763, 1976.
21. Lane, D. A., Thomson, B. A., Lovett, A. M., and Reid, N. M., Real-time tracking of industrial emissions through populated areas using a mobile APCI mass spectrometer system, *Adv. Mass Spectrom.*, 8, 1480, 1980.
22. Siegel, N. W. and McKeown, M. C., Ions and electrons in the electron capture detector. Quantitative detection by atmospheric pressure ionization mass spectrometry, *J. Chromatogr.*, 122, 397, 1976.

23. Siegel, M. W. and Fite, W. L., Terminal ions in weak atmospheric pressure plasmas. Applications of atmospheric pressure ionization to trace impurity analysis in gases, *J. Phys. Chem.*, 80, 2871, 1976.
24. Dzidic, I., Carroll, D. I., Stillwell, R. N., Horning, M. G., and Horning, E. C., Studies of negative ions by atmospheric pressure ionization mass spectrometry, *Adv. Mass Spectrom.*, 7, 319, 1978.
25. Lawson, G., Bonner, R. F., and Todd, J. F. J., The quadrupole ion store (quistor) as a novel source for a mass spectrometer, *J. Phys. E:* 6, 357, 1973.
26. Herod, A. A. and Harrison, A. G., Bimolecular reactions of ions trapped in an electron space charge, *Int. J. Mass Spectrom. Ion Phys.*, 4, 415, 1970.
27. Lehman, T. A. and Bursey, M. M., *Ion Cyclotron Resonance Spectrometry,* John Wiley & Sons, New York, 1976.
28. Henis, J. M. S., An ion cyclotron resonance study of the ion-molecule reactions in methanol, *J. Am. Chem. Soc.*, 90, 844, 1968.
29. Henis, J. M. S., Analytical implications of ion cyclotron resonance spectrometry, *Anal. Chem.*, 41(10), 22A, 1969.
30. Gross, M. L. and Wilkins, C. L., Ion cyclotron resonance spectrometry: recent advances of analytical interest, *Anal. Chem.*, 43(14), 65A, 1971.
31. Hunter, R. L. and McIver, R. T., Conceptual and experimental basis for rapid scan ion cyclotron resonance spectroscopy, *Chem. Phys. Lett.*, 49, 577, 1977.
32. Hunter, R. L. and McIver, R. T., Mechanism of low-pressure chemical ionization mass spectrometry, *Anal. Chem.*, 51, 699, 1979.
33. Comisarow, M. B. and Marshall, A. G., Fourier transform ion cyclotron resonance spectroscopy, *Chem. Phys. Lett.*, 25, 282, 1974.
34. Ghaderi, S., Kulkarni, P. S., Ledford, E. B., Wilkins, C. L., and Gross, M. L., Chemical ionization Fourier transform mass spectrometry, *Anal. Chem.*, 53, 428, 1981.
35. Jennings, K. R., Negative chemical ionization mass spectrometry, in *Mass Spectrometry,* Spec. Period. Rep., Vol. 4, Chemical Society, London, 1977, chap. 9.
36. Hunt, D. F., Stafford, G. C., Crow, F. W., and Russell, J. W., Pulsed positive negative ion chemical ionization mass spectrometry, *Anal. Chem.*, 48, 2098, 1976.
37. McFadden, W. H., *Technique of Combined Gas Chromatography/Mass Spectrometry: Applications in Organic Analysis,* Interscience, New York, 1973.
38. Blum, W. and Richter, W. J., Analyses of alkene mixtures by combined capillary gas chromatography-chemical ionization mass spectrometry, *Tetrahedron Lett.*, 835, 1973.
39. Arsenault, G. P., Dolhun, J. J., and Biemann, K., Gas chromatography-chemical ionization mass spectrometry, *Chem. Commun.*, 1542, 1970.
40. Schoengold, D. M. and Munson, B., Combination of gas chromatography and chemical ionization mass spectrometry, *Anal. Chem.*, 43, 1811, 1970.
41. Oswald, E. O., Fishbein, L., Corbett, B. J., and Whaler, M. P., Metabolism of naturally occurring propenyl benzene derivatives. II. Separation and identification of tertiary aminopropiophenones by combined gas-liquid chromatography and chemical ionization mass spectrometry, *J. Chromatogr.*, 73, 43, 1972.
42. Hatch, F. and Munson, B., Techniques in gas chromatography/chemical ionization mass spectrometry, *Anal. Chem.*, 49, 169, 1977.
43. Arpino, P. J. and Guiochon, G., LC/MS coupling, *Anal. Chem.*, 51, 682A, 1979.
44. Arpino, P. J., Baldwin, M. A., and McLafferty, F. W., Liquid chromatography-mass spectrometry. II. Continuous monitoring, *Biomed. Mass Spectrom.*, 1, 80, 1974.
45. McFadden, W. H., Schwartz, H. L., and Evans, S., Direct analysis of liquid chromatographic effluents, *J. Chromatogr.*, 122, 389, 1976.
46. Craig, R. D., Bateman, R. H., Green, B. N., and Millington, D. S., Mass spectrometry instrumentation for chemists and biologists, *Phil. Trans. R. Soc. London,* A293, 135, 1979.
47. Beuhler, R. J., Flanigan, E., Greene, L. J., and Friedman, L., Proton transfer mass spectrometry of peptides. A rapid heating technique for underivatized peptides containing arginine, *J. Am. Chem. Soc.*, 96, 3990, 1974.
48. Cotter, R. J., Laser desorption chemical ionization mass spectrometry, *Anal. Chem.*, 52, 1767, 1980.
49. Cotter, R. J., Mass spectrometry of non-volatile compounds: desorption from extended probes, *Anal. Chem.*, 52, 1589A, 1980.
50. Baldwin, M. A. and McLafferty, F. W., Direct chemical ionization of relatively involatile samples. Application to underivatized oligopeptides, *Org. Mass Spectrom.*, 7, 1353, 1973.
51. Hunt, D. F., Shabanowitz, J., Botz, F. K., and Brent, D. A., Chemical ionization mass spectrometry of salts and thermally labile organics with field desorption emitters as solids probes, *Anal. Chem.*, 49, 1161, 1977.
52. Hansen, G. and Munson, B., Surface chemical ionization mass spectrometry, *Anal. Chem.*, 50, 1130, 1978.

53. Cotter, R. J., Probe for direct exposure of solid samples to the reagent gas in a chemical ionization mass spectrometer, *Anal. Chem.*, 51, 317, 1979.
54. Carroll, D. I., Dzidic, I., Horning, M. G., Montgomery, F. E., Nowlin, J. G., Stillwell, R. N., Thenot, J.-P., and Horning, E. C., Chemical ionization mass spectrometry of non-volatile organic compounds, *Anal. Chem.*, 51, 1858, 1979.
55. Hansen, G. and Munson, B., Chemical ionization mass spectrometry of thermally labile compounds, *Anal. Chem.*, 52, 245, 1980.
56. Bruins, A. P., Probe for the introduction of non-volatile solids into a chemical ionization source by thermal desorption from a platinum wire, *Anal. Chem.*, 52, 605, 1980.
57. Kondrat, R. W. and Cooks, R. G., Direct analysis of mixtures by mass spectrometry, *Anal. Chem.*, 50, 81A, 1978.
58. Yost, R. A. and Enke, C. G., Triple quadrupole mass spectrometry for direct mixture analysis and structure elucidation, *Anal. Chem.*, 51, 1251A, 1978.
59. McLafferty, F. W., Tandem mass spectrometry (MS/MS): a promixing new analytical technique for specific component determination in complex mixture, *Accts. Chem. Res.*, 13, 33, 1980.
60. Cooks, R. G. and Glish, G. L., Mass spectrometry/mass spectrometry, *Chem. Eng. News*, 59(48), 40, 1981.
61. Boyd, R. K. and Beynon, J. H., Scanning of sector mass spectrometers to observe the fragmentation of metastable ions, *Org. Mass Spectrom.*, 12, 163, 1977.
62. Bruins, A. P., Jennings, K. R., and Evans, S., The observation of metastable transitions in a double-focussing mass spectrometer using a linked scan of the electric sector and magnetic sector fields, *Int. J. Mass Spectrom. Ion Phys.*, 26, 395, 1978.
63. Weston, A. F., Jennings, K. R., Evans, S., and Elliott, R. M., The observation of metastable transitions in a double-focussing mass spectrometer using a linked scan of the accelerating and electric-sector voltages, *Int. J. Mass Spectrom. Ion Phys.*, 20, 317, 1976.
64. Kemp, D. L., Cooks, R. G., and Beynon, J. H., Simulated MIKE spectra from a conventional double focussing mass spectrometer, *Int. J. Mass Spectrom. Ion Phys.*, 21, 93, 1976.
65. Yost, R. A. and Enke, C. G., Selected ion fragmentation with a tandem quadrupole mass spectrometer, *J. Am. Chem. Soc.*, 100, 2274, 1978.
66. Yost, R. A., Enke, C. G., McGilvery, D. C., Smith, D., and Morrison, J. D., High efficiency collision-induced dissociation in an RF-only quadrupole, *Int. J. Mass Spectrom. Ion Phys.*, 30, 127, 1979.
67. Hunt, D. F., Shabanowitz, J., and Giordani, A. B., Collision activated decomposition of negative ions in mixture analysis with a triple quadrupole mass spectrometer, *Anal. Chem.*, 52, 386, 1980.
68. Lacey, M. J. and Macdonald, C. G., Constant neutral spectrum in mass spectrometry, *Anal. Chem.*, 51, 691, 1979.

Chapter 4

CHEMICAL IONIZATION REAGENT ION SYSTEMS

I. INTRODUCTION

A wide variety of chemical ionization reagent ion systems have been investigated to date. The following discussion reviews the major systems used in both positive and negative ion chemical ionization and outlines the major types of ionization reactions observed. Detailed considerations of the CI mass spectra of various classes of compounds are deferred to Chapter 5.

II. POSITIVE ION REAGENT GAS SYSTEMS

A. Brønsted Acid Reagent Systems

By far the most widely used reagent gas systems have been those which yield Brønsted acids (BH^+); these react primarily by proton transfer to species of higher proton affinity than B. Table 1 provides a list of reagent gases and the major reactant ions (BH^+) produced in these gases along with the proton affinities of the conjugate bases B. The H_3^+ ion, formed by Reaction 1, is the sole product observed in hydrogen at high pressures.

$$H_2^{+\cdot} + H_2 \rightarrow H_3^+ + H^\cdot \tag{1}$$

Reaction 1 is exothermic by $\sim$41 kcal mol^{-1}, consequently the H_3^+ formed initially is highly excited and is deactivated by collision with the H_2 reagent gas.[1,2] Unless this deactivation process is complete the exothermicity of proton transfer from H_3^+ will be greater than predicted from proton affinity data. In 10% X/90% H_2 mixtures (X = N_2, CO_2, N_2O, CO) the XH^+ ion dominates at pressures of 0.5 to 1.0 torr.[3,4] The XH^+ ions are formed by Reactions 1 plus 2 to 4, all of which are known to occur quite rapidly.[5-7]

$$H_3^+ + X \quad \rightarrow XH^+ + H_2 \tag{2}$$

$$H_2^{+\cdot} + X \quad \rightarrow XH^+ + H^\cdot \tag{3}$$

$$X^{+\cdot} + H_2 \quad \rightarrow XH^+ + H^\cdot \tag{4}$$

In the N_2O/H_2 mixture significant ion signals (10 to 15%) are observed for NO^+; this ion is formed by electron impact ionization of N_2O and does not react with H_2. Methane, the most widely used reagent gas and that used in the initial CI study,[8] produces the most complicated spectrum at high pressures. At 1 torr total pressure the major ions and their relative abundances are CH_5^+ (48%), $C_2H_5^+$ (41%) and $C_3H_5^+$ (6%).[9] There are also minor yields of $C_2H_4^+$ and $C_3H_7^+$. The major ions formed in methane on electron impact are $CH_4^{+\cdot}$, CH_3^+, and CH_2^+. These react by Reactions 5 to 9 to yield the major ions observed.

$$CH_4^{+\cdot} + CH_4 \rightarrow CH_5^+ + CH_3^\cdot \tag{5}$$

$$CH_3^+ + CH_4 \rightarrow C_2H_5^+ + H_2 \tag{6}$$

$$CH_2^{+\cdot} + CH_4 \rightarrow C_2H_4^{+\cdot} + H_2 \tag{7}$$

$$\rightarrow C_2H_3^+ + H_2 + H^\cdot \tag{8}$$

$$C_2H_3^+ + CH_4 \rightarrow C_3H_5^+ + H_2 \tag{9}$$

Table 1
BRØNSTED ACID CHEMICAL IONIZATION
REAGENTS

Reagent gas	Reactant ion (BH+)	PA(B)[a]	HIA(BH+)[a]	RE(BH+)[b]
H_2	H_3^+	100.7	300	9.3
N_2/H_2	N_2H^+	117.4	283	8.6
CO_2/H_2	CO_2H^+	128.6	272	8.1
N_2O/H_2	N_2OH^{+c}	137.0	264	7.7
CO/H_2	HCO^+	141.4	259	7.5
CH_4	CH_5^+	130.5	270	8.0
	$C_2H_5^+$	163.5	272	8.4
H_2O	$H^+(H_2O)_n$[d]	173.0	227	6.2
CH_3OH	$H^+(CH_3OH)_n$[d]	184.9	119	6.0
C_3H_8	$C_3H_7^+$	184.9	250	7.5
$i-C_4H_{10}$	$C_4H_9^+$	196.9	231	6.9
NH_3	$H^+(NH_3)_n$[d]	205.0	195	4.8

[a]   kcal mol^{-1}.

[b]   eV.

[c] NO^+ also observed.

[d]   Degree of solvation depends on partial pressure of reagent gas; thermochemical data for monosolvated proton.

Propane has not seen significant use as a chemical ionization reagent ion. The dominant ion in propane at high pressure is $C_3H_7^+$ with minor yields of $C_3H_6^{+\cdot}$ and $C_3H_8^{+\cdot}$;[10,11] the primary ions formed by electron impact on C_3H_8 react with propane primarily by H^- abstraction. In isobutane at high pressures $C_4H_9^+$, presumed to be the t-butyl ion, is the dominant ion (>92%) with a minor yield of $C_3H_7^+$ (~3%);[12] again the primary fragment ions react with $i-C_4H_{10}$ by H^- abstraction.[10] The polar reagent gases H_2O, CH_3OH, and NH_3 produce the solvated proton $H^+(B)_n$ with the extent of solvation depending on the total pressure; typically at 1 torr pressure ions with n up to 3 to 4 are observed.[13] To decrease the extent of solvation it is possible to dilute the polar reagent gas with a nonpolar reagent gas such as methane or, for ammonia, isobutane. The $CH_3OH^{+\cdot}$, CH_2OH^+, $CH_2O^{+\cdot}$ and CHO^+ primary ions formed in methanol all react relatively rapidly to form $CH_3OH_2^+$[14] which subsequently form more highly solvated species. In both H_2O and NH_3 the following reaction sequence[15] involving reaction of the primary ions $XH_2^{+\cdot}$ and XH^+ (X = O, NH) lead to formation of XH_3^+ which then forms the higher solvated species

$$XH_2^{+\cdot} + XH_2 \rightarrow XH_3^+ + XH^\cdot \tag{10}$$

$$XH^+ + XH_2 \rightarrow XH_2^{+\cdot} + XH \tag{11}$$

$$\rightarrow XH_3^+ + X \tag{12}$$

In atmospheric pressure chemical ionization using ambient air carrier gas the dominant positive reactant ions are solvated protons $H^+(H_2O)_n$ with the range of n depending on the partial pressure of water in the ambient air.[16] The major primary ions produced in air will be $N_2^{+\cdot}$ and $O_2^{+\cdot}$. Kebarle and co-workers[17,18] have elucidated the mechanism by which these primary ions are converted to $H^+(H_2O)_n$. For $N_2^{+\cdot}$ the reaction sequence is

$$N_2^{+\cdot} + 2N_2 \rightarrow N_4^{+\cdot} + N_2 \tag{13}$$

$$N_4^{+\cdot} + H_2O \rightarrow H_2O^{+\cdot} + 2N_2 \tag{14}$$

$$H_2O^{+\cdot} + H_2O \rightarrow H_3O^+ + OH \tag{15}$$

followed by clustering. Neither $O_2^{+\cdot}$ nor $O_4^{+\cdot}$ are capable of undergoing charge transfer to H_2O and the conversion proceeds mainly as follows

$$O_2^{+\cdot} + 2O_2 \qquad \rightarrow O_4^{+\cdot} \tag{16}$$

$$O_2^{+\cdot} + H_2O + O_2 \quad \rightarrow O_2^{+\cdot}(H_2O) + O_2 \tag{17}$$

$$O_4^{+\cdot} + H_2O \qquad \rightarrow O_2^{+\cdot}(H_2O) + O_2 \tag{18}$$

$$O_2^{+\cdot}(H_2O) + H_2O \rightarrow H_3O^+ \cdot OH + O_2 \tag{19}$$

$$H_3O^+ \cdot OH + H_2O \rightarrow H^+(H_2O)_2 + OH^{\cdot} \tag{20}$$

The major reaction mode of the Brønsted acids, BH^+, is proton transfer (Reaction 21) to give initially the protonated molecule MH^+.

$$BH^+ + M \rightarrow MH^+ + B \tag{21}$$

As the data in Chapter 2, Section III.B. show this reaction usually occurs with high efficiency provided it is exothermic, i.e., provided PA(M)->PA(B). Examination of the proton affinity data (Tables 17 to 20, Chapter 2) indicates that $C_3H_7^+$ and reagent ions higher in Table 1 than $C_3H_7^+$ will react by exothermic proton transfer to most organic molecules. By contrast $C_4H_9^+$ and $H^+(NH_3)_n$ are more specific protonating agents, particularly the latter ion. In general $H^+(NH_3)_n$ will protonate organic compounds with PA>205 kcal mol^{-1}; in practice this means that primarily nitrogen-containing compounds will be protonated. With compounds having PA<205 kcal mol^{-1} the ammonia CI mass spectra frequently show $[M + NH_4]^+$ adduct ions. In a systematic study Keough and De Stefano[19] showed that $[M + NH_4]^+$ was not formed to a significant extent if the proton affinity of M was less than $\sim$188 kcal mol^{-1}; it also appeared that a lone pair of electrons at the basic site is a necessary prerequisite for formation of $[M + NH_4]^+$. In line with these conclusions it has been reported[13] that $H^+(NH_3)_n$ is unreactive with alkanes, alkenes, aromatics, simple alcohols, ethers, and nitro compounds. Other cluster ions frequently observed (usually in low abundance) are $[M + C_2H_5]^+$ and $[M + C_3H_5]^+$ in CH_4 CI mass spectra, $[M + C_3H_7]^+$ in propane CI mass spectra, and $[M + C_3H_3]^+$ and $[M + C_4H_9]^+$ in i-butane CI mass spectra; when polar reagent gases are used low intensity $[MH^+ + B]$ species frequently are observed.

In principle the BH^+ ions also can react by hydride ion abstraction (Reaction 22) or by charge exchange (Reaction 23). Accordingly we have given in Table 1 the effective

$$BH^+ + M \rightarrow [M - H]^+ + BH_2 \text{ (or } B + H_2) \tag{22}$$

$$BH^+ + M \rightarrow M^{+\cdot} + BH^{\cdot} \text{ (or } B + H^{\cdot}) \tag{23}$$

hydride ion affinities (HIAs) and recombination energies (REs) of the BH^+ reactant ions. The HIAs should be considered in light of the values given in Table 21, Chapter 2. The H_3^+ ion has a high HIA and in most H_2CI mass spectra appreciable $[M—H]^+$ ion signals are observed. In addition ground state H_3^+ has a recombination energy of $\sim$9.3 eV and is capable of undergoing charge exchange with any molecule having an ionization energy lower than this value; consequently $M^{+\cdot}$ ions are seen in many H_2 CI mass spectra. The N_2H^+ ion is a moderately strong hydride ion abstracting reagent and is capable of undergoing charge exchange with species having ionization energies below $\sim$8.6 eV. The CO_2H^+, N_2OH^+, and HCO^+ reagent ions are weaker hydride ion abstracting reagents but still are capable of abstracting secondary and tertiary alkyl hydrogens, allylic and benzylic hydrogens, and those alpha to hydroxyl groups. These reactant ions are unlikely to react to any significant extent by charge exchange. It should be noted that 10 to 15% of the ion yield in N_2O/H_2 mixtures resides in NO^+ and the characteristic reaction of this ion (Section II.E.) are also observed when using this reagent gas system.

The CH_5^+ and $C_2H_5^+$ ions produced in methane are capable of abstracting hydride from many classes of molecules and $[M-H]^+$ ions frequently are observed in methane CI mass spectra. The $C_2H_5^+$ ion has RE $\approx$ 8.5 eV and will charge exchange only with molecules of quite low ionization energy. However, the methane reagent ion system includes a low-abundance $C_2H_4^{+\cdot}$ ion (RE = 9 to 11 eV[20]) which can charge exchange with many species to produce low-abundance (usually $\leqslant$1%) molecular ions of the additive. The $C_3H_5^+$ ion is capable of abstracting hydride ion from a variety of molecules but the remaining reagent ions are very weak hydride ion abstracting reagents; they also have low recombination energies and are unlikely to react by charge exchange.

The exothermicity of the proton transfer Reaction 21 resides largely in the $M-H^+$ bond formed[3] and is available as internal energy to promote fragmentation of the MH^+ ion. To maximize the probability of MH^+ ion formation, which provides molecular weight information, one utilizes the least exothermic protonation reaction possible. On the other hand, structural information is derived from the fragment ions observed in the CI mass spectrum. The extent of fragmentation depends on the identity of M and the exothermicity of the protonation reaction. Both effects are illustrated by the proton transfer CI mass spectra shown in Figures 1 and 2. Clearly the extent of fragmentation of each compound decreases and the MH^+ ion intensity increases as the protonation exothermicity decreases. Equally clear is that the oxime shows a much greater tendency for the MH^+ ion to fragment. It frequently is observed that CH_4 CI mass spectra combine an adequate MH^+ ion abundance for molecular weight determination with appropriate fragmentation for structure elucidation; consequently CH_4 is the most useful single reagent gas.

The fragmentation reactions of the even-electron MH^+ ion usually involve elimination of an even-electron neutral (a stable molecule) to form an even-electron fragment ion. If the molecule contains a functional group Y the initial fragmentation frequently involves loss of the stable molecule HY from MH^+; note as an example the loss of H_2O and CH_3OH from MH^+ in the spectra of Figure 2. The tendency of the functional group Y to be eliminated as HY is inversely related to the proton affinity of HY.[23] This can be rationalized by noting that the reverse of Reaction 24 is the addition of

$$RY \cdot H^+ \rightarrow R^+ + HY \qquad (24)$$

R^+ to HY. If R^+ were a proton the exothermicity of the reverse reaction would be the proton affinity of HY. Thus it is reasonable to project that the endothermicity (and, hence, the critical reaction energy) of the forward reaction should be proportional to PA(HY) and that as PA(HY) decreases a greater fraction of the RYH^+ will fragment. Accordingly one predicts that the order of the ease of loss of Y as HY should be

$$NH_2 < CH_3C(:O)O < CH_3S < CH_3O < C_6H_5 \approx HC(:O)O < CN < HS < HO < I \approx Cl \approx Br$$

There is some experimental support for this inverse dependence of the extent of fragmentation of $RY \cdot H^+$ on PA(HY).[24,25] Table 2 shows the $MH^+/C_6H_{11}^+$ intensity ratios observed[24] in the CH_4 CI mass spectra of cyclohexyl derivatives ($C_6H_{11}Y$). There is a general correlation of the ratio with PA(HY) although when other fragmentation routes are important, for example the formation of protonated acid ions from cyclohexyl esters, the correlation begins to break down.

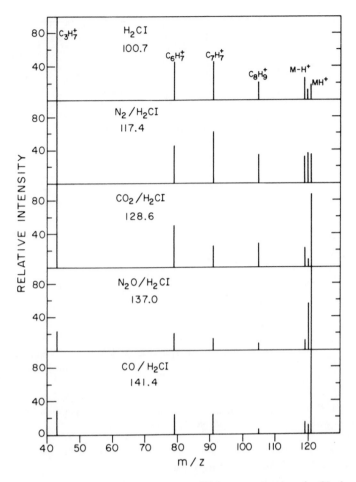

FIGURE 1. Effect of reagent proton affinity on proton transfer CI of n-propyl benzene (PA in kcal mol⁻¹). Data from Reference 21.

While this approach suffices to rationalize the relative extent of fragmentation of related monofunctional compounds the problem of predicting the fragmentation of more complex, polyfunctional, compounds is much more difficult. One approach which has been used frequently is to assume that protonation at a given site or functionality triggers fragmentation at that site. As an example, it was assumed[26] in early studies of the CH_4 CI of amino acids that protonation at the hydroxyl group resulted in elimination of water, protonation at the carbonyl oxygen led to elimination of formic acid, while protonation at the amino group led to elimination of ammonia (Scheme 1).

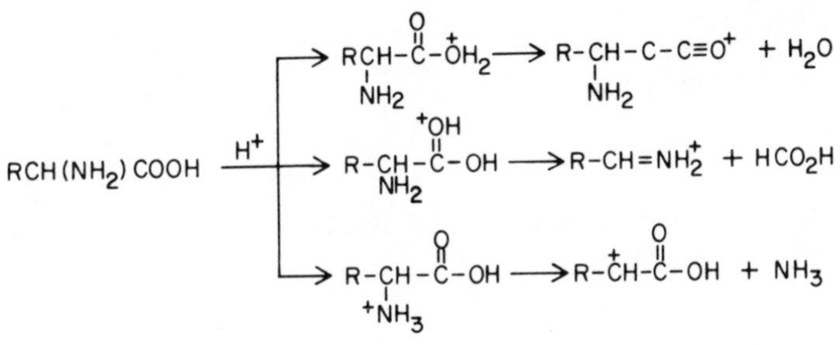

SCHEME 1

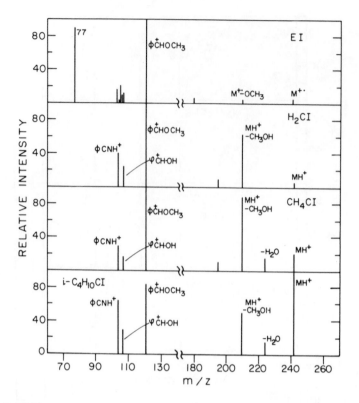

FIGURE 2.   Effect of protonation exothermicity on CI mass spectrum of benzoin methyl ether oxime. Data from Reference 22.

More recent studies[27] have shown that the added proton interchanges between the various basic sites prior to fragmentation of MH^+, casting doubt on this localized activation model.

An alternative approach, suggested by detailed studies of several classes of compounds[3,4,21,28,29] is to assume a competitive kinetic scheme for fragmentation of MH^+, similar to the quasi-equilibrium theory used to rationalize electron impact mass spectra.[30,31] In this approach the initial protonation reaction is assumed to produce a collection of MH^+ ions with an average excess internal energy determined by the exothermicity of the protonation reaction. These excited MH^{+*} ions are assumed to be sufficiently long-lived that the excess energy is randomized among the internal degrees of freedom. The MH^{+*} ions may undergo collisional de-excitation (Reaction 25), or alternatively, may fragment by one of several possible competing fragmentation reactions (Reactions 26 and 27). For each fragmentation reaction the rate coefficient $k(E)$

$$MH^{+*} \xrightarrow{B} MH^+ \qquad (25)$$

$$MH^{+*} \longrightarrow F_1^+ + N_1 \qquad (26)$$

$$\longrightarrow F_2^+ + N_2 \qquad (27)$$

can be calculated by suitable application of absolute rate theory and, in its simplest form, can be approximated by[30,31]

$$k(E) = v\left(\frac{E-\epsilon_o}{E}\right)^s \qquad (28)$$

Table 2
FRAGMENTATION OF PROTONATED
CYCLOHEXYL DERIVATIVES $c\text{-}C_6H_{11}Y \cdot H^+ \rightarrow$
$c\text{-}C_6H_{11}^+ + HY$

Y	PA(HY)[a]	$MH^+/C_6H_{11}^+$	Other ions[b]
SCOCH$_3$	—	1.7	$CH_3COSH_2^+$ (73%)
NH$_2$	205	1.7	$[M-H]^+$ (70%)
NCO	—	1.3	—
PH$_2$	—	1.0	$[M-H]^+$ (38%)
NCS	—	0.65	—
C$_6$H$_5$	182	0.23	$[M-H]^+$ (23%)
OCOC$_2$H$_5$	193	0.17	$C_2H_5CO_2H_2^+$ (310%)
OCOCH$_3$	191	0.15	$CH_3CO_2H_2^+$ (84%)
SH	176	0.14	—
OCOH	183	0.07	$HCO_2H_2^+$ (13%)
OH	173	0.03	—
F,Cl,Br,I	~140	0.00	$[M-H]^+$ (1—7%)

[a] kcal mol^{-1}.
[b] Expressed as % MH$^+$ or $C_6H_{11}^+$ whichever is larger.

where v is an energy-independent term or frequency factor, E is the excess internal energy of the fragmenting ion, ε_o is the critical energy for fragmentation (often called the activation energy), and s is the effective member of oscillators, often taken as ~1/3 the total number (3N-6). At the usual pressures of the bath gas (0.5 to 1.0 torr) collisional de-excitation is relatively rapid and fragmentation reactions must have rate coefficients greater than ~10^7 sec^{-1} for fragmentation to compete with stabilization. (This minimum rate coefficient is greater than the minimum value of ~10^5 sec^{-1} applicable in electron impact-induced fragmentation.)

The factors determining reaction competition have been discussed elsewhere[30,31] for electron impact-induced fragmentations; the basic conclusions are applicable here. In general, relative rate coefficients for competing fragmentation reactions are determined primarily by the relative critical energies at low excess internal energies and by the relative frequency factors at high internal energies. The frequency factor is lower for a fragmentation process involving rearrangement than it is for a simple bond-fission reaction. It appears that the relative critical energies, which frequently can be related to reaction enthalpies, play a major role in CI-induced fragmentation reactions. Thus, in many cases, the relative importance of fragmentation reactions can be deduced if the relative endothermicities for fragmentation of ground state MH$^+$ can be estimated. Table 3 illustrates this for the two major competing fragmentation reactions (Reactions 29 and 30) for protonated butyl acetates.[28]

$$CH_3CO_2C_4H_9 \cdot H^+ \rightarrow CH_3C(OH)_2^+ + C_4H_8 \qquad (29)$$

$$\rightarrow C_4H_9^+ + CH_3CO_2H \qquad (30)$$

For t-butyl acetate formation of t-C$_4$H$_9^+$ is thermochemically favored and C$_4$H$_9^+$ is the dominant fragment ion. In contrast for s-butyl acetate formation of CH$_3$C(OH)$_2^+$ is thermochemically favored and this species forms the base peak in the spectrum. The results for i-butylacetate can be rationalized only if it is assumed that there is a rearrangement to the t-butyl$^+$ structure during fragmentation. Such rearrangements of alkyl ions during fragmentation appear to be relatively common.[32] As the average internal energy is increased by increasing the protonation exothermicity, relative frequency

Table 3
RELATIVE FRAGMENT ION INTENSITIES
IN CO_2/H_2 CI OF BUTYL ACETATES
CH_3CO_2R

R	Rel. int. $CH_3C(OH)_2^+$	$\Delta H^{\circ a}$	Rel. int. $C_4H_9^+$	$\Delta H^{\circ a}$
n-C_4H_9	100	25	31	47
				$(33)^b$
i-C_4H_9	89	25	100	47
				$(21)^c$
s-C_4H_9	100	25	22	33
t-C_4H_9	9	29	100	21

a ΔH° for reaction of ground state $CH_3CO_2R \cdot H^+$, kcal mol^{-1}.

b   Value in bracket for s-$C_4H_9^+$ structure.

c   Value in bracket for t-$C_4H_9^+$ structure.

factors should play an increasing role in determining relative rates. The rearrangement reaction (Reaction 29) should have a lower frequency factor than the simple bond-fission reaction (Reaction 30). Thus the $C_4H_9^+/CH_3C(OH)_2^+$ ratio in s-butyl acetate is observed to be 0.13 in the CO/H_2 CI mass spectrum but increases to 0.62 in the more exothermic H_2 CI system.

Even when the reaction thermochemistry cannot be calculated explicitly, the thermochemically favored reaction frequently can be deduced from a knowledge of the factors which determine relative ion and neutral stabilities. Two examples will suffice as illustrations. The ion (Structure A) formed by elimination of the elements of formic acid from protonated amino acids is greatly stabilized by the amino group and constitutes the most abundant fragment ion in the CH_4 CI mass spectra of many amino acids.[26,27] In contrast the [MH$^+$−NH$_3$] ion (Structure B) is destabilized by the adjacent carbonyl group and loss of NH$_3$ is observed only when the critical reaction energy is lowered by anchimeric assistance by a neighboring group as in the formation of Structure C from protonated methionine

$$R-\overset{+}{C}H-NH_2 \rightleftharpoons RCH=\overset{+}{N}H_2 \qquad\qquad R-\underset{+}{CH}-\overset{\overset{\displaystyle O}{\|}}{C}-OH \qquad\qquad \begin{matrix} CH_2-\overset{+}{S}-CH_3 \\ | \quad\quad | \\ CH_2-CH-COOH \end{matrix}$$

STRUCTURE A   STRUCTURE B   STRUCTURE C

As a second example it is noted that considerations of the relative proton affinities of H_2O and CH_3OH would suggest that H_2O loss should be preferred over CH_3OH loss from the protonated oxime of Figure 2. However the benzylic ion formed by CH_3OH loss is more stable than the ion formed by loss of H_2O and [MH$^+$−CH$_3$OH] is the more abundant ion in the CI mass spectra.

Finally it should be pointed out that while formation of an even-electron ion by elimination of an even-electron neutral from MH$^+$ normally is the thermochemically favored fragmentation route, there are a few examples where formation of an odd-electron fragment ion by elimination of an odd-electron neutral (radical) is the favored mode of fragmentation. For example, the CH_4 CI mass spectra of iodobenzene and the iodotoluenes show[33] abundant ion signals corresponding to [MH$^+$−I$^\cdot$] and no ion signal corresponding to [MH$^+$−HI]. Similarly the CH_4 CI mass spectra of the bromoanisoles and bromoanilines show[34] extensive fragmentation to form [MH$^+$−Br$^\cdot$]

with little formation of [MH$^+$–HBr]. In all cases the high heat of formation of the appropriate phenyl cation relative to the appropriate odd-electron benzene molecular ion makes Reaction 31 energetically more demanding than Reaction 32.

$$RC_6H_4 X \cdot H^+ \rightarrow RC_6H_4^+ + HX \tag{31}$$

$$\rightarrow RC_6H_5^{+\cdot} + X^\cdot \tag{32}$$

B. Hydride Ion Abstraction Reagent Systems

There have been no specific reagent systems developed which involve hydride ion abstraction as the major or sole ionization reaction. The NO$^+$ ion reacts with some classes of compounds primarily by H$^-$ abstraction but also shows other reaction modes; the use of NO as a reagent gas is discussed in Section II.E. The C$_2$H$_5^+$ ion produced in methane reacts with alkanes entirely by H$^-$ abstraction,[35] but usually reacts predominantly by proton transfer when the proton affinity of the substrate molecule makes proton transfer exothermic. The data in Table 21, Chapter 2, shows that CF$_3^+$ has a high hydride ion affinity while the data summarized in Table 6, Chapter 2, show that CF$_3^+$ reacts moderately rapidly with alkanes by H$^-$ abstraction; these results suggest that CF$_3^+$ might make an efficient H$^-$ abstraction reagent ion. The preliminary results reported[36] do not bear out this prediction.

C. Charge Exchange Reagent Systems

An alternative mode of ionization which has seen some use is that of charge exchange, Reaction 33.

$$R^{+\cdot} + M \rightarrow M^{+\cdot} + R \tag{33}$$

In this mode the odd-electron M$^{+\cdot}$ molecular ion, characteristic of electron impact ionization, is the initial product ion. Consequently the fragmentation reactions which are observed are the same as those observed in EI mass spectrometry. The essential difference is that in electron impact the M$^{+\cdot}$ ions have a distribution of internal energies whereas in charge exchange the M$^{+\cdot}$ ions have a discrete internal energy determined by the exothermicity of Reaction 33. Consequently the electron impact mass spectrum is the net result of summing an energy-dependent competitive kinetic scheme over a distribution of internal energies, while the charge exchange mass spectrum is the net result of the energy-dependent kinetic scheme at a fixed internal energy determined by the charge exchange exothermicity. This exothermicity is determined by the recombination energy of the reactant ion less the ionization energy of the additive M. Table 4 provides a partial list of possible charge exchange reagent systems with their recombination energies. Most organic molecules have ionization energies in the 8 to 12 eV range. The effect of varying the reaction exothermicity on the charge exchange mass spectrum is shown by the data in Table 5 for cyclohexane (IE = 9.88 eV).[37] The results clearly show the distinct difference between EI and charge exchange mass spectra and the strong dependence of the charge exchange mass spectrum on the recombination energy of the reactant ion.

The rare gas ions, N$_2^{+\cdot}$, CO$_2^{+\cdot}$, CO$^{+\cdot}$, and C$_6$H$_6^{+\cdot}$ can be produced by electron impact ionization of the appropriate pure gas at high source pressures; at such pressures some formation of dimer ions will occur. These dimer ions will have recombination energies which are lower than the monomer ions by the binding energy in the dimeric species. For example, this binding energy is ∼1 eV for the N$_2^+ \cdot$N$_2$ dimers.[38] The CS$_2^{+\cdot}$ ion has been produced[39-41] by electron impact ionization of N$_2$/CS$_2$ mixture, where at high pressures CS$_2^{+\cdot}$ is the major (>80%) product ion; minor yields of S$_2^{+\cdot}$, (CS$_2$)$_2^{+\cdot}$ and CS$_3^+$

Table 4
CHARGE EXCHANGE REAGENT SYSTEMS

Reagent gas	Reactant ion	Recombination energy (eV)
C_6H_6	$C_6H_6^{+\cdot}$	9.3
N_2/CS_2	$CS_2^{+\cdot}$	~10.0
CO/COS	$COS^{+\cdot}$	11.2
Xe	$Xe^{+\cdot}$	12.1, 13.4
CO_2	$CO_2^{+\cdot}$	13.8
CO	$CO^{+\cdot}$	14.0
Kr	$Kr^{+\cdot}$	14.0, 14.7
N_2	$N_2^{+\cdot}$	15.3
Ar	$Ar^{+\cdot}$	15.8, 15.9

Table 5
ELECTRON IMPACT AND CHARGE EXCHANGE MASS SPECTRA OF CYCLOHEXANE

Ionization	Fractional intensities						
	$C_6H_{12}^{+\cdot}$	$C_5H_9^+$	$C_5H_8^{+\cdot}$	$C_4H_7^{+\cdot}$	$C_3H_7^+$	$C_3H_6^{+\cdot}$	$C_3H_5^+$
EI (70 eV)	0.26	0.07	0.31	0.10	0.04	0.08	0.15
$CE(CS_2^{+\cdot})$	1.0	—	—	—	—	—	—
$CE(COS^{+\cdot})$	0.91	0.03	0.06	—	—	—	—
$CE(Xe^{+\cdot})$	0.03	0.15	0.06	0.11	—	0.10	0.01
$CE(CO^{+\cdot})$	0.08	0.05	0.41	0.22	—	0.08	0.15
$CE(N_2^{+\cdot})$	0.06	0.01	0.12	0.27	—	0.07	0.49

are observed.[41] Similarly, $COS^{+\cdot}$ has been produced[41] by ionization of a COS/CO mixture at high pressures; minor yields of $S_2^{+\cdot}$, $CS_2^{+\cdot}$, and $(COS)_2^{+\cdot}$ also were observed.

Charge exchange has not seen extensive use as a chemical ionization method. Since the fragmentation reactions following charge exchange ionization are the same as those following electron impact ionization, structure elucidation capabilities are unlikely to be enhanced by charge exchange ionization. The potential uses of charge exchange as a method of ionization would appear to be twofold. In mixtures with proton transfer reagents (Brønsted acids), which through formation of MH^+ provide molecular weight information, charge exchange can be utilized to produce fragment ions to assist in structure elucidation. Alternatively, low-energy charge exchange can be used to simplify mass spectra, in the limit producing only $M^{+\cdot}$, without the reduction in sensitivity inherent in the use of low-energy electron impact ionization; in suitable cases low-energy charge exchange can be used to selectively ionize the components in mixtures which have low ionization energies.

In the first use, Hunt et al.[13,42] have shown that Ar/H_2O reagent gas mixtures give good MH^+ ion intensities for a variety of compounds through reaction of $H^+(H_2O)_n$ as well as fragment ions, from reaction of $Ar^{+\cdot}$, which are characteristic of molecular structure. In the second use it has been shown that N_2/NO mixtures give NO^+ (recombination energy 8.3 to 9.3 eV) as the dominant reactant ion and that the abundance of molecular ions are considerably enhanced through use of this reagent system.[43-45] However, as discussed in Section II.E., NO^+ reacts by a number of different reaction channels with different molecules and care must be taken in interpreting the results obtained with this reagent gas. Subba Rao and Fenselau[46] have used $C_6H_6^{+\cdot}$ (produced in benzene) for selective ionization of esters of unsaturated fatty acids in the presence of saturated fatty acid esters. A similar use of $C_6H_6^{+\cdot}$ for selective ionization of aro-

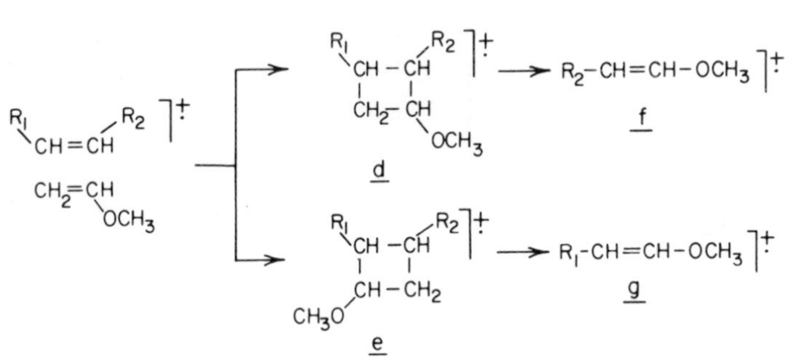

$$R_1 \diagdown CH = CH \diagup R_2 \Big]^{+\cdot}$$

$$CH_2 = CH \diagdown OCH_3$$

$$\longrightarrow \quad R_1 \diagdown CH - CH \diagup R_2 \Big]^{+} \quad \longrightarrow \quad R_2-CH=CH-OCH_3 \Big]^{+\cdot}$$

$$CH_2-CH \diagdown OCH_3$$

d   **f**

$$\longrightarrow \quad R_1 \diagdown CH - CH \diagup R_2 \Big]^{+} \quad \longrightarrow \quad R_1-CH=CH-OCH_3 \Big]^{+\cdot}$$

$$CH_3O \diagdown CH - CH_2$$

e   **g**

SCHEME 2

matic hydrocarbons in hydrocarbon mixtures has been suggested by Sunner and Szabo.[47] Charge exchange from $C_6H_6^{+\cdot}$ also has been used for selective ionization of polychlorinated biphenyls (PCBs) in atmospheric pressure chemical ionization experiments.[48] Similarly, Sieck[49] has used $C_6H_{12}^{+\cdot}$ (produced by photoionization of cyclohexane) to ionize selectively aromatic hydrocarbons in gasolines and fuel oils. At the other extreme Ryan[50] has used high-energy charge exchange to remove interferences by dissociative ionization when monitoring polycyclic aromatic hydrocarbons (PAHs) in complex mixtures. The PAHs formed stable molecular ions while the interfering materials were dissociatively ionized to lower mass, noninterfering fragment ions when ionized by charge exchange with $Ar^{+\cdot}$.

D. Condensation Reaction Reagents

Condensation ion/molecule reactions have not found extensive applications in chemical ionization mass spectrometry. A method of locating olefinic double bonds without derivatization which involves a condensation ion/molecule reaction between vinyl methyl ether (VME) and the olefinic compound has been developed by Jennings and co-workers.[51,52] The principle involved is illustrated in Scheme 2. Four-centered addition complexes (Structures D and E) are formed by attack of either olefinic molecular ion on the second neutral olefin; these fragment, in part, to characteristic ionic products in Structures F and G, the m/z ratios of which identify the size of R_1 and R_2 and, hence, the position of the double bond. Alternative fragmentation reactions of the complexes involve loss of $CH_3^{\cdot}$ and loss of CH_3OH. It is not known which olefinic moiety is ionized and quite likely both molecular ions are capable of reacting with the other neutral.

In the early studies[51,52] either pure vinyl methyl ether (VME) or VME/CO_2 mixtures were used as the reagent gas. Ionization of pure VME produces a complex array of ions as does ionization of the ~10% VME/90% CO_2 mixtures where charge exchange from the primary ions $CO_2^{+\cdot}$ to either VME or the added olefin results in extensive fragmentation. Considerably simpler spectra are obtained if a reagent gas mixture containing approximately 75% N_2, 20% CS_2, and 5% VME is used.[53,54] In this latter mixture the initial electron impact process produces mainly $N_2^{+\cdot}$ which by charge exchange forms $CS_2^{+\cdot}$. The $CS_2^{+\cdot}$ ion is a low-energy charge exchange reagent ion (Table 4) and on reaction with either vinyl methyl ether or the olefinic compound produces predominantly molecular ions. Figure 3 compares the high-pressure spectra of N_2/VME and N_2/CS_2/VME mixtures. This method of double bond location has been shown to be useful in locating such bonds in olefins, unconjugated diolefins, and olefinic fatty acid methyl esters.[53,54]

E. NO as a Positive Ion Reagent System

Electron impact on either pure NO or N_2/NO mixtures at high pressure produces NO^+ in good yield with lesser amount of the dimeric ion $(NO)_2^{+\cdot}$. The chemical ioniza-

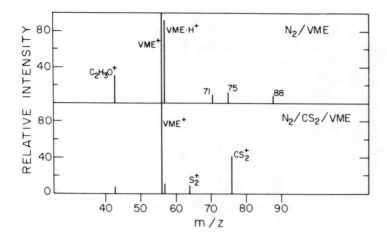

FIGURE 3.   Mass spectra of N_2/vinyl methyl ether and N_2/CS$_2$/vinyl methyl ether mixtures at high pressures. Data from Reference 54.

tion reactions of the NO$^+$ ion have received considerable study.[13,43-45,55-60] Three alternative reaction channels, addition (Reaction 34), hydride ion abstraction (Reaction 35), or charge exchange (Reaction 36) are observed depending on the molecule.

$$\text{NO}^+ + \text{M} \;\rightarrow\; \text{M}\cdot\text{NO}^+ \tag{34}$$

$$\rightarrow\; [\text{M} - \text{H}]^+ + \text{HNO} \tag{35}$$

$$\rightarrow\; \text{M}^{+\cdot} + \text{NO}\cdot \tag{36}$$

In addition, in some cases hydroxide ion abstraction from alcohols or halide ion abstraction from alkyl halides are observed. The ionization energy of NO is 9.3 eV,[61] but it has been observed[56] that with aromatic molecules significant charge exchange is observed only when the IE of the aromatic is below ~8.5 eV. It appears that the effective recombination energy of NO$^+$ is in the range 8.5 to 9.0 eV and charge exchange can be expected with molecules with ionization energies in this range or lower.

The hydride ion affinity of NO$^+$ is ~246 kcal mol^{-1} with the result that NO$^+$ is capable of abstracting, as hydride ion, secondary or tertiary alkyl hydrogens, secondary allylic hydrogens, benzylic hydrogens, and hydrogens bonded to the same carbon atom as hydroxyl groups (see Table 21, Chapter 2). In general, NO$^+$ reacts with primary and secondary alcohols by hydride ion abstraction, with tertiary alcohols by OH$^-$ abstraction, with ketones and esters by NO$^+$ addition, with aldehydes by H$^-$ abstraction and NO$^+$ addition, and with ethers by H$^-$ abstraction. With internal olefins NO$^+$ reacts by hydride ion abstraction and by NO$^+$ addition or charge exchange depending on the ionization energy of the olefin. With terminal olefins, an additional reaction mode involves addition of NO$^+$ to the olefin followed by elimination of small neutral alkene molecules to give a series of ions $C_nC_{2n}NO^+$ ($n = 4,5,6$).

There are several problems associated with the use of NO as a reagent gas. It has been reported[57] that the lifetime of a rhenium filament in the presence of 1 torr of NO is ~6 hr. In addition, in several cases there are oxidation reactions; the mechanisms of which are not entirely clear but presumably involve neutral species. Thus primary alcohols show [M−3H]$^+$ and [M−2H + NO]$^+$ ions in addition to [M−H]$^+$; apparently the alcohol is oxidized to an aldehyde which reacts in the typical manner. Secondary alcohols apparently oxidize to ketones and show [M−2H + NO]$^+$ ion signals in addition to [M − H]$^+$ ion signals. While these product ions give some indication of alcohol structure they also overlap with the products expected from aldehydes and ketones, making analysis of mixtures of alcohols and carbonyl compounds impossible. Similarly alkanes in addition to [M−H]⁺ ions show [M−3H]⁺ and [M−2H + NO]⁺ ions resulting from apparent oxidation of the alkane to an alkene. These products overlap with the products expected from the NO CI of alkenes making analysis of mixtures of alkanes and alkenes impossible.

F. Miscellaneous Positive Ion Reagent Systems

A number of polar reagent gases have been used to produce proton-bound adducts, MHX^+; these include X = dimethylamine with a sulfone,[62] X = trimethylamine with nucleosides,[63] and ethanolamine and ethylenediamine with a bifunctional oxygen compound.[64] In addition, Fenselau et al.[65,66] have observed that NH_4^+ does not form stable adducts (or significant MH^+ ion intensities) in the NH_3 CI of glucoronides, but that addition of pyridine to the reagent gas results in $MH^+ \cdot C_5H_5N$ adduct ions which provide molecular weight information. Tetramethylsilane at high pressure gives predominantly $Si(CH_3)_3^+$ which reacts with compounds containing n or π electrons to give [M + $Si(CH_3)_3$]⁺ adduct ions.[67]

Under ICR conditions biacetyl gives primarily $(CH_3CO)_2^+$ which reacts as a protonating and acetylating agent with a variety of ketones, esters, nitriles, and nitro compounds.[68-72] Under high-pressure CI conditions it was necessary to dilute the biacetyl with helium to obtain useful CI spectra;[73] even then the reagent gas mass spectrum was complex and it was necessary to employ computer subtraction of spectra to obtain useful CI mass spectra. Some stereoselectivity was noted in the importance of the acetylation reaction.

A number of ICR studies of the reactions of metal ions with organic substrates have been reported.[74-79] A rich chemistry is involved which may have eventual applications in chemical ionization studies. Of particular interest is the generation of metal cations by pulsed laser ionization from a solid source which avoids the necessity of operating at high source pressures.[80]

III. NEGATIVE ION REAGENT GAS SYSTEMS

A. Electron Capture Reagent Systems

As discussed in Section IV.A., Chapter 2, the capture of electrons by polyatomic molecules is a resonance process which normally requires electrons of near-thermal energy. Production of a large number of electrons in a restricted low energy range at low pressures is a difficult experimental task; as a result negative ion mass spectrometry at low pressures has not been utilized to nearly the same extent as positive ion mass spectrometry. Despite this problem considerable understanding of the negative ion mass spectra of a wide variety of compounds has been obtained from low-pressure studies.[81,82]

Under CI operating conditions the simplest type of process which leads to a negative ion mass spectrum is that in which the reagent gas acts only as a moderating gas to produce a high population of near-thermal energy electrons which are captured by the sample molecules. The incident electron beam is thermalized by inelastic electron/molecule interactions (including ionization of the molecule); in addition the electrons produced in the ionization reactions also are thermalized. In early experiments[83] nitrogen was used as the moderating gas; methane or isobutane at pressures near 1 torr now are commonly used.[84,85] Any oxygen-containing impurities present in the reagent gas or present as background in the mass spectrometer may produce low concentrations of H⁻, OH⁻, or O⁻ ions; reaction of these ions can make the CI spectrum obtained unreproducible as well as more complicated to interpret. To minimize the problems of lack of reproducibility Hass and co-workers[86,87] have deliberately added a small amount of O_2 to the methane moderating gas and obtained so-called oxygen-enhanced methane negative ion CI mass spectra. In addition to improving reproducibility, more structural differentiation was obtained in a number of cases. In the CH_4/O_2 mixtures one has near-thermal energy electrons and, possibly some O_2^- ions; as a consequence the final spectrum is the result not only of electron capture but also reactions of O_2^- with the neutral and reactions of the ions formed by electron capture with neutral O_2. The oxygen added has deleterious effects on hot filaments[87,88] and oxygen as a reagent gas is best used with a Townsend discharge ion source.[89]

In pulsed positive and negative chemical ionization (PPNCI) mass spectrometry[85] the use of N_2 as the moderating gas leads to N_2^+ as a charge exchange reagent gas in the positive ion mode and to near-thermal energy electrons in the negative ion mode. The use of methane or isobutane as the moderating gas leads to the appropriate Brønsted acids in the positive ion mode and to near-thermal energy electrons in the negative ion mode.

The main usefulness of electron capture chemical ionization lies in its potential for molecular weight identification through molecular anion formation and in the possibility of enhanced sensitivity compared to other ionization reactions owing to the higher rates of electron capture reactions (Section IV.A., Chapter 2). Not all molecules will have the necessary properties to achieve these goals; thus, like electron capture detection in gas chromatography, electron capture CI will not be universally applicable. However, in many cases it is possible to derivatize the molecule to give it a positive electron affinity necessary for formation of M^-, and a high rate coefficient for electron capture, necessary for enhanced sensitivity. Examples of such derivatizations include the use of pentafluorobenzaldehyde to form Schiff bases with aromatic amines, pentafluorobenzoyl chloride for reaction with phenols and amines, and tetrafluorophthalic anhydride for reaction with amines.[85,90]

B. Brønsted Base Reagent Systems

In negative ion CI mass spectrometry Brønsted bases, reacting predominantly by Reaction 37, play a role analogous to the role frequently played by Brønsted acids in

$$B^- + M \rightarrow BH + [M - H]^- \quad \Delta H = \Delta H_{acid}(M) - \Delta H_{acid}(BH)$$

$$= PA ([M - H]^-) - PA(B^-)$$

$$(37)$$

positive ion CI mass spectrometry. As discussed in Section IV.D., Chapter 2, Reaction 37 is likely to be rapid for simple B^- provided it is exothermic. Thus the reaction should be efficient provided BH is a weaker acid than M or, equivalently, provided PA(B^-) is greater than PA([M−H]$^-$). Since most of the exothermicity of Equation 37 is likely to reside in the B−H bond formed, little fragmentation of [M−H]$^-$ is expected. Table 6 provides a list of proton affinities for a number of negative ions which are candidate Brønsted base reagent ions. Also included in the table are the electron affinities of the neutrals B which are relevant to the tendency of the ion to react by charge exchange. The H$^-$ ion does not appear to have been used as a chemical ionization reagent ion; the NH$_2^-$ ion, produced by electron capture in NH$_3$, has seen only very limited use.[91] The O$^-$ and O$_2^-$ ions react by several routes other than proton abstraction and will be discussed separately in later sections.

The OH$^-$ ion has been produced by electron impact on $N_2O/He/H_2$ or $N_2O/He/CH_4$ mixtures in 1:1:1 ratios.[92] The reactions occurring are

Table 6
BRØNSTED BASE
REAGENT SYSTEMS

B⁻	PA(B⁻)[a]	EA(B)[a]
H⁻	400	17.4
NH₂⁻	400	18.0
OH⁻	382	42.2
O⁻̇	382	33.7
CH₃O⁻	379	36.2
F⁻	372	78.4
O₂⁻̇	351	10.1
Cl⁻	333	83.4

[a] kcal mol⁻¹.

$$e_{therm} + N_2O \rightarrow O^{-\cdot} + N_2 \tag{38}$$

$$O^{-\cdot} + \begin{Bmatrix} H_2 \\ CH_4 \end{Bmatrix} \rightarrow OH^- + \begin{Bmatrix} H^\cdot \\ CH_3^\cdot \end{Bmatrix} \tag{39}$$

At 3 torr total source pressure OH⁻ constituted ~93% of the total ionization in the N₂O/He/H₂ mixture, with the major impurity being (H₂O)OH⁻ (3%). By contrast, in the N₂O/He/CH₄ mixture at 3 torr total pressure OH⁻ was only ~72% of the total ionization with the major impurity (~18%) being an ion of m/z 26 (possibly CN⁻). At lower source pressures some near-thermal energy electrons may still be present and, if electron capture is very rapid, electron capture ionization also may be observed.[93] The OH⁻ ion is a quite strong base and is capable of abstracting a proton from a wide range of organic molecules; at the same time the electron affinity is quite high and it is unlikely to react by charge exchange except in rare cases.

The reactions of OH⁻ with a variety of organic molecules under CI conditions have been investigated.[92-97] With carboxylic acids, alcohols, ketones, and amino acids the [M−H]⁻ ion is formed in good yields with a minor yield of [M−3H]⁺ ions being observed in some systems; these presumably arise by elimination of H₂ from the [M−H]⁻ ion. Esters normally show [M−H]⁻ ions as well as significant ion signals for the carboxylate anion RCOO⁻. The few ethers and aliphatic amines examined did not show [M−H]⁻ ions but produced more complicated spectra, particularly the amines. Alkylbenzenes and olefins react to produce [M−H]⁻ initially; these ions are reactive with the N₂O reagent gas giving [M+43]⁻ ([M−H+N₂O]⁻) and [M+25]⁻ ([M−H + N₂O−H₂O]⁻).

The usefulness of CH₃O⁻ as a CI reagent ion has not been investigated systematically. It is almost as strong a base as OH⁻ and should react readily with many organic compounds by proton abstraction; the studies to date confirm this.[85] With its high electron affinity CH₃O⁻ is unlikely to react by charge exchange. The CH₃O⁻ ion is readily prepared by adding about 1% CH₃ONO to a moderating gas of CH₄.[85] In the pulsed positive ion-negative ion mode of operation the CH₄/CH₃ONO mixture gives the Brønsted acids CH₅⁺ and C₂H₅⁺ in the positive ion mode and the Brønsted base CH₃O⁻ in the negative ion mode. Other weaker bases B⁻ can be obtained[85] by adding BH to the CH₄/CH₃ONO mixture until the CH₃O⁻ signal disappears.

The chemical ionization reactions of F⁻ with a variety of molecules have been investigated by Tiernan and co-workers.[98] They found that electron impact on CHF₃ at 0.1

torr predominantly produced clusters of the type $(HF)_nF^-$ (n = 1 to 3) with practically no F^-, while electron impact on NF_3 at 0.1 torr produced largely F^- with only a minor yield of $(HF)F^-$; the latter system was used in the studies reported. The F^- ion is a weaker Brønsted base than OH^- or CH_3O^- and is not capable of abstracting a proton from alkanols, amines, alkanes, alkenes, or most nitriles. With the simple alkanols investigated, a low yield of $[M + F]^-$ was observed, however the major product observed was $(HF)F^-$. With cycloalkane diols F^- is reported to react by both proton abstraction to form $[M-H]^-$ and formation of $[M + F]^-$,[99] with the relative intensities depending on the stereochemistry of the diol. Amines gave only $(HF)F^-$, CN^-, and an unidentified product of m/z 42. Alkanes gave a very low yield of $[M + F]^-$ while alkenes and alkylbenzenes were unreactive. With the exception of acetonitrile, nitriles gave no $[M-H]^-$ ions the major reaction product being CN^- presumably produced by the S_N2 displacement reaction

$$F^- + RCN \rightarrow RF + CN^- \tag{40}$$

Similarly, esters gave predominantly the corresponding carboxylate anions and ether gave alkoxide ions, presumably in both cases the result of displacement reactions. On the other hand carboxylic acids, aldehydes, and ketones gave predominantly $[M-H]^-$ with low yields of $[M + F]^-$. It does not appear that F^- will be a particularly useful CI reagent ion.

The Cl^- ion has been produced under CI conditions by dissociative electron capture by CH_2Cl_2,[84,100] by $CHCl_3$,[101,102] and by CF_2Cl_2.[103] The Cl^- ion is a weak Brønsted base and will abstract a proton only from very acidic compounds. It has been used in atmospheric pressure ionization experiments to monitor diphenylhydantoin and barbiturates through formation of the $[M-H]^-$ ion.[101,102] With a series of organophosphorus compounds of general formula $(RO)_2PO(CH_2)_nR'$ the major ion observed was $(RO)_2PO^-$; this product was attributed to electron capture[104] but may equally well arise by an S_N2 displacement reaction of Cl^-. The major reaction of Cl^- is to form the attachment ion $[M + Cl]^-$. The intensity of the attachment product appears to be related to the acidity of the substrate. Thus carboxylic acids, amides, amino acids, aromatic amines, and phenols give relatively intense attachment ions while aliphatic alcohols and amines, carbonyl compounds, ethers, nitroaromatics, and chlorinated hydrocarbons give lower yields of attachment ions. Aliphatic and aromatic hydrocarbons, tertiary amines, and nitriles give no attachment ions. Chloride attachment has been used to identify saccharides, where useful structural information also is obtained from fragment ions[103] and to monitor pentachlorophenol residues in environmental samples.[105]

C. O⁻ As a Reagent Ion

Electron capture by nitrous oxide (N_2O), using a nonreactive moderating gas such as N_2 or a rare gas, is a convenient source of O^-.[106] The reactions of O^- with a variety of organic molecules have been investigated by Jennings and co-workers using both ICR techniques[107-109] and high pressure chemical ionization techniques.[106,110-112] In addition the O^- CI of a number of alcohols have been reported by Houriet et al.[97] A number of different types of reactions are observed; the relative importance of these reactions depends strongly on the structure of the substrate molecule. These reactions may be categorized as follows:

Hydrogen Atom Abstraction:

$$O^- + M \rightarrow OH^- + [M - H]^\bullet \tag{41}$$

The OH⁻ ion is one of the products of the reaction of O⁻ with most organic molecules (see Reaction 39 for the preparation of OH⁻). The OH⁻ ion may subsequently react as a Brønsted base with the additive.

Proton Abstraction:

$$O^{\cdot-} + M \rightarrow OH^{\cdot} + [M - H]^- \tag{42}$$

The O⁻ ion is a relatively strong Brønsted base and is capable of abstracting a proton from many organic substrates. Frequently the [M−H]⁻ ion is the most abundant high-mass ion found; it may originate in part by reaction of OH⁻ as well. The [M−H]⁻ ion may undergo fragmentation as in the formation of [M−3H]⁻ in the O⁻ CI mass spectra of alcohols.[97]

H₂⁺ Abstraction:

$$O^{\cdot-} + M \rightarrow H_2O + [M - 2H]^{\cdot-} \tag{43}$$

If the organic molecule contains a suitably activated CH_2 group $H_2^{\cdot}$ abstraction may occur. The reaction was first observed with C_2H_4 where deuterium labeling showed that both hydrogens originated from the same carbon.[107] Subsequent work has shown that the reaction occurs with many 1-alkenes,[108] primary alkyl nitriles,[108] carbonyl compounds,[109] and a number of aromatic compounds.[106] Although the results for ethylene indicate that the two hydrogens come from the same carbon, this is not always the case. Thus in $H_2^{\cdot}$ abstraction from m-xylene one H originates from each methyl group,[106] while in $H_2^{\cdot}$ abstraction from propyne the acetylenic hydrogen is involved to some extent.[113] In $H_2^{\cdot}$ abstraction from acetone 1,3-$H_2^{\cdot}$ abstraction occurs about 56% of the time.[114] In a number of cases the [M−2H]⁻ ion undergoes further fragmentation as, for example, in Reaction 44 for ketones:

$$\underset{\substack{| \\ \cdot}}{R-\overset{\overset{O}{\|}}{C}-C}-R' \rightarrow R'-C \equiv C-O^- + R^{\cdot} \tag{44}$$

H-Atom Displacement:

$$O^{\cdot-} + M \rightarrow [M - H + O]^- + H^{\cdot} \tag{45}$$

Almost all aromatic compounds investigated give a peak at [M + 15]⁻ corresponding to the phenoxide anion derived by displacement of a hydrogen atom of the aromatic ring. Similarly, aldehydes react in part by H-atom displacement giving the relevant carboxylate anion.

Alkyl Group Displacement:

$$O^{\cdot-} + M \rightarrow [M - R + O]^- + R^{\cdot} \tag{46}$$

Displacement of alkyl groups leading to the appropriate carboxylate anion is observed in the reaction of O⁻ with carbonyl compounds.[109] As examples, reaction of O⁻ with $C_3H_7COCH_3$ produces both $C_3H_7COO^-$ and CH_3COO^- while reaction with the isomeric $(C_2H_5)_2CO$ produces only $C_2H_5COO^-$ by alkyl group displacement. Methyl-substituted aromatic compounds show low-intensity [M + 1]⁻ ion signals arising by displacement of the methyl group by O⁻.

D. O_2^- as a Reagent Ion

A mixture of hydrogen and oxygen in an electron impact ion source has been used as a source of O_2^{-}[85] while a Townsend discharge in pure O_2 at 1 torr pressure also has been used to produce O_2^- as a reagent ion.[90] Negative ion CI mass spectra using CH_4/O_2 mixtures also have been measured;[86-88] these may reflect, in part, reactions of O_2^-. In these relatively low-pressure ($\sim$1 torr) experiments near-thermal energy electrons also are likely to be present since the lifetime with respect to autodetachment of the O_2^-* formed by electron capture is very short (Table 10, Chapter 2), and collisional stabilization very inefficient. At the much higher pressures of atmospheric pressure chemical ionization collisional stabilization will be more effective and O_2^-, and hydrated forms thereof, are the major negative reactant ions using air[16,115] or N_2/O_2 mixtures[116] as carrier gases. The lifetimes of hot filaments are seriously reduced in contact with O_2; in addition when rhenium filaments are used background ions corresponding to the anions of rhenium trioxide and rhenium tetroxide have been reported.[88] Townsend discharges or atmospheric pressure ionization clearly are preferable when using O_2 as a reagent gas.

O_2^- is a stronger Brønsted base than Cl^- and will react by proton abstraction with a wider range of substrates. For example, it will abstract a proton from all three isomeric nitrophenols but not from the weaker acids phenol and p-chlorophenol.[115] The electron affinity of O_2^- is low ($\sim$10 kcal mol^{-1}) and it also can react by charge exchange with molecules of suitably high electron affinity.[85,90] Additional reaction modes are clustering to form $[M + O_2]^-$,[85,90] displacement of H to form $[M-H+O]^-$ (Reaction 47),[90] or displacement of Cl to form $[M-Cl+O]^-$ (Reaction 48).[115]

$$O_2^- + M \rightarrow [M - H + O]^- + OH^{\cdot} \tag{47}$$

$$O_2^- + M \rightarrow [M - Cl + O]^- + OCl^{\cdot} \tag{48}$$

In many of these reactions, as well as more complex reactions such as that forming $C_6H_2Cl_2O_2^{-\cdot}$ in the O_2 CI of tetrachlorodibenzodioxins,[116,117] it is not clear whether the reaction is that of O_2^- with the substrate, reaction of M^- with neutral O_2, or both reactions. This is particularly true at the lower pressures where significant populations of near-thermal energy electrons are present to produce M^- by electron attachment.

REFERENCES

1. Leventhal, J. J. and Friedman, L., Energy transfer in the de-excitation of $(H_3^+)^*$ by H_2, *J. Chem. Phys.*, 50, 2928, 1969.
2. Bowers, M. T., Chesnavich, W. J., and Huntress, W. T., Deactivation of internally excited H_3^+ ions. Comparison of experimental product distributions of reactions of H_3^+ with CH_3NH_2, CH_3OH, and CH_3SH with predictions of quasi-equilibrium theory calculations, *Int. J. Mass Spectrom. Ion Phys.*, 12, 357, 1973.
3. Bowen, R. D. and Harrison, A. G., Chemical ionization mass spectra of selected C_3H_6O compounds, *Org. Mass Spectrom.*, 16, 159, 1981.
4. Herman, J. A. and Harrison, A. G., Effect of reaction exothermicity on the proton transfer chemical ionization mass spectra of isomeric C_5 and C_6 alkanols, *Can. J. Chem.*, 59, 2125, 1981.

5. Shannon, T. W. and Harrison, A. G., Concurrent ion-molecule reactions. II. Reactions in X-D$_2$ mixtures, *J. Chem. Phys.*, 43, 4206, 1965.

6. Harrison, A. G. and Myher, J. J., Ion-molecule reactions in mixtures with D$_2$ or CD$_4$, *J. Chem. Phys.*, 46, 3276, 1967.

7. Albritton, D. L., Ion-neutral reaction rate constants measured in flow reactors through 1977, *At. Data Nucl. Data Tables*, 22, 1, 1978.

8. Munson, M. S. B. and Field, F. H., Chemical ionization mass spectrometry. I. General introduction, *J. Am. Chem. Soc.*, 88, 2621, 1966.

9. Field, F. H. and Munson, M. S. B., Reactions of gaseous ions. XIV. Mass spectrometric studies of methane at pressures to 2 torr, *J. Am. Chem. Soc.*, 87, 3289, 1965.

10. Munson, M. S. B., Franklin, J. L., and Field, F. H., High pressure mass spectrometric study of alkanes, *J. Phys. Chem.*, 68, 3098, 1964.

11. Solka, B. H., Lau, A. Y.-K., and Harrison, A. G., Bimolecular reactions of trapped ions. VIII. Reactions in propane and methane-propane mixtures, *Can. J. Chem.*, 52, 1798, 1974.

12. Field, F. H., Chemical ionization mass spectrometry. IX. Temperature and pressure studies with benzyl acetate and t-amyl acetate, *J. Am. Chem. Soc.*, 91, 2829, 1969.

13. Hunt, D. F., Reagent gases for chemical ionization mass spectrometry, *Adv. Mass Spectrom.*, 6, 517, 1974.

14. Gupta, S. K., Jones, E. G., Harrison, A. G., and Myher, J. J., Reactions of thermal energy ions. VI. Hydrogen transfer ion-molecule reactions involving polar molecules, *Can. J. Chem.*, 45, 3107, 1967.

15. Huntress, W. T. and Pinizzotto, R. F., Product distributions and rate constants for ion-molecule reactions in water, hydrogen sulfide, ammonia, and methane, *J. Chem. Phys.*, 59, 4742, 1973.

16. Lane, D. A., Thomson, B. A., Lovett, A. M., and Reid, N. M., Real time tracking of industrial emissions through populated areas using a mobile APCI mass spectrometer system, *Adv. Mass Spectrom.*, 8, 1480, 1980.

17. Good, A., Durden, D. A., and Kebarle, P., Ion-molecule reactions in pure nitrogen and nitrogen containing traces of water at total pressures of 0.5-4.0 torr. Kinetics of clustering reactions forming H$^+$(H$_2$O)$_n$, *J. Chem. Phys.*, 52, 212, 1970.

18. Good, A., Durden, D. A., and Kebarle, P., Mechanism and rate constants of ion-molecule reactions leading to formation of H$^+$(H$_2$O)$_n$ in moist oxygen and air, *J. Chem. Phys.*, 52, 222, 1970.

19. Keough, T. and DeStefano, A. J., Factors affecting reactivity in ammonia chemical ionization mass spectrometry, *Org. Mass Spectrom.*, 16, 537, 1981.

20. Lindholm, E., Mass spectra and appearance potentials studied by use of charge exchange in a tandem mass spectrometer, in *Ion-Molecule Reactions*, Vol. 2, Franklin, J. L., Ed., Plenum Press, New York, 1972, chap. 10.

21. Herman, J. A. and Harrison, A. G., Effect of protonation exothermicity on the chemical ionization mass spectra of some alkylbenzenes, *Org. Mass Spectrom.*, 16, 423, 1981.

22. Harrison, A. G. and Kallury, R. K. M. R., Stereochemical applications of mass spectrometry. I. The utility of EI and CI mass spectrometry in differentiation of isomeric benzoin oximes and phenylhydrazones, *Org. Mass Spectrom.*, 15, 249, 1980.

23. Field, F. H., Chemical ionization mass spectrometry in *Mass Spectrometry*, Ser. 1, Vol. 5, MTP Rev. Sci., Phys. Chem., Maccoll, A., Ed., Butterworth, London, 1972, chap. 5.

24. Jardine, I. and Fenselau, C., Proton localization in chemical ionization fragmentation, *J. Am. Chem. Soc.*, 98, 5086, 1976.

25. Harrison, A. G. and Onuska, F. I., Fragmentation in chemical ionization mass spectrometry and the proton affinity of the departing neutral, *Org. Mass Spectrom.*, 13, 35, 1978.

26. Milne, G. W. A., Axenrod, T., and Fales, H. M., Chemical ionization mass spectrometry of complex molecules. IV. Amino acids, *J. Am. Chem. Soc.*, 92, 5170, 1970.

27. Tsang, C. W. and Harrison, A. G., The chemical ionization of amino acids, *J. Am. Chem. Soc.*, 98, 1301, 1976.

28. Herman, J. A. and Harrison, A. G., Structural and energetics effects in the fragmentation of protonated esters in the gas phase, *Can. J. Chem.*, 59, 2133, 1981.

29. Bowen, R. D. and Harrison, A. G., Chemical ionization mass spectra of some C$_4$H$_8$O isomers, *J. Chem. Soc. Perkin Trans. II*, 1544, 1981.

30. Tsang, C. W. and Harrison, A. G., The origin of mass spectra, in *Biochemical Applications of Mass Spectrometry*, Waller, G. R., Ed., John Wiley & Sons, New York, 1972, 135.

31. Williams, D. H. and Howe, I., *Principles of Organic Mass Spectrometry*, McGraw-Hill, New York, 1972.

32. Schwarz, H. and Stahl, D., Unimolecular water loss from protonated alcohols in the gas phase: the effect of ion/dipole interactions on the isomerization of the incipient carbocation, *Int. J. Mass Spectrom. Ion Phys.*, 36, 285, 1980.

33. Leung, H.-W. and Harrison, A. G., Structural and energetics effects in the chemical ionization of halogen-substituted benzenes and toluenes, *Can. J. Chem.*, 54, 3439, 1976.

34. Leung, H.-W. and Harrison, A. G., Specific substituent effects in the dehalogenation of halobenzene derivatives by the gaseous Brønsted acid CH_5^+, *J. Am. Chem. Soc.*, 102, 1623, 1980.

35. Houriet, R., Parisod, G., and Gaümann, T., The mechanism of chemical ionization of n-paraffins, *J. Am. Chem. Soc.*, 99, 3599, 1977.

36. Rudat, M. A., Chemical ionization with CF_4, in 29th Annu. Conf. Mass Spectrometry and Allied Topics, Minneapolis, May 24 to 29, 1981.

37. Herman, J. A., Li, Y.-H., and Harrison, A. G., Energy dependence of the fragmentation of some $C_6H_{12}^+$ ions, *Org. Mass Spectrom.*, 17, 143, 1982.

38. Kebarle, P., Higher order reactions-ions clusters and ion solvation, in *Ion-Molecule Reactions*, Franklin, J. L., Ed., Plenum Press, New York, 1972.

39. Meot-Ner, M. and Field, F. H., Thermodynamic ionization energies and charge transfer reactions in benzene and substituted benzenes, *Chem. Phys. Lett.*, 44, 484, 1976.

40. Meot-Ner, M., Hamlet, P., and Field, F. H., Bonding energies in association ions of aromatic compounds. Correlations with ionization energies, *J. Am. Chem. Soc.*, 100, 5466, 1978.

41. Li, Y.-H., Herman, J. A., and Harrison, A. G., Charge exchange mass spectra of some C_8H_{10} isomers, *Can. J. Chem.*, 59, 1753, 1981.

42. Hunt, D. F. and Ryan, J. F., Argon-water mixtures as reagents for chemical ionization mass spectrometry, *Anal. Chem.*, 44, 1306, 1972.

43. Jelus, B., Munson, B., and Fenselau, C., Charge exchange mass spectra of trimethylsilyl ethers of biologically important compounds: an analytical technique, *Anal. Chem.*, 46, 729, 1974.

44. Jelus, B., Munson, B., and Fenselau, C., Reagent gases for GC-MS analysis, *Biomed. Mass Spectrom.*, 1, 96, 1974.

45. Hunt, D. F. and Harvey, T. M., Nitric oxide chemical ionization mass spectra of alkanes, *Anal. Chem.*, 47, 1965, 1975.

46. Subba Rao, S. C. and Fenselau, C., Evaluation of benzene as a charge exchange reagent, *Anal. Chem.*, 50, 511, 1978.

47. Sunner, J. and Szabo, I., The analytical use of chemical ionization in a tandem mass spectrometer, *Adv. Mass Spectrom.*, 7, 1383, 1978.

48. Thomson, B. A., Sakuma, T., Fulford, J., Lane, D. A., Reid, N. M., and French, J. B., Fast *in situ* measurement of PCB levels in ambient air to ng m^{-3} levels using a mobile atmospheric pressure chemical ionization mass spectrometer system, *Adv. Mass Spectrom.*, 8, 1422, 1980.

49. Sieck, L. W., Finger printing and partial quantification of complex hydrocarbon mixtures by chemical ionization mass spectrometry, *Anal. Chem.*, 51, 128, 1979.

50. Ryan, P. W., Some advantages of argon as a chemical ionization gas in analytical GC/MS applications, in 25th Annu. Conf. Mass Spectrometry and Allied Topics, Washington, D.C., May 29 to June 3, 1977.

51. Ferrer-Correia, A. J., Jennings, K. R., and Sen Sharma, D. K., The use of ion-molecule reactions in the mass spectrometric location of double bonds, *Org. Mass Spectrom.*, 11, 867, 1976.

52. Ferrer-Correia, A. J., Jennings, K. R., and Sen Sharma, D. K., The use of ion-molecule reactions in the mass spectrometric location of double bonds, *Adv. Mass Spectrom.*, 7, 287, 1978.

53. Greathead, R. J. and Jennings, K. R., The location of double bonds in mono- and di-unsaturated compounds, *Org. Mass Spectrom.*, 15, 431, 1980.

54. Chai, R. and Harrison, A. G., Location of double bonds by chemical ionization mass spectrometry, *Anal. Chem.*, 53, 34, 1981.

55. Hunt, D. F. and Ryan, J. F., Chemical ionization mass spectrometry studies: nitric oxide as a reagent gas, *J. Chem. Soc. Chem. Commun.*, 620, 1972.

56. Einolf, N. and Munson, B., High pressure charge exchange mass spectrometry, *Int. J. Mass Spectrom. Ion Phys.*, 9, 141, 1972.

57. Hunt, D. F. and Harvey, T. M., Nitric oxide chemical ionization mass spectra of olefins, *Anal. Chem.*, 47, 2136, 1975.

58. Jardine, I. and Fenselau, C., Charge exchange mass spectra of morphine and tropane alkaloids, *Anal. Chem.*, 47, 730, 1975.

59. Jardine, I. and Fenselau, C., The high pressure nitric oxide mass spectra of aldehydes, *Org. Mass Spectrom.*, 10, 748, 1975.

60. Munson, B., Chemical ionization mass spectrometry-analytical applications of ion-molecule reactions, in *Interactions Between Ions and Molecules*, Ausloos, P., Ed., Plenum Press, New York, 1975, 505.

61. Rosenstock, H. M., Draxl, K., Steiner, B. W., and Herron, J. T., Energetics of gaseous ions, *J. Phys. Chem. Ref. Data*, Suppl. 1, 6, 1977.

62. Vouros, P. and Carpino, L. A., Selective chemical ionization mass spectrometry as an aid in the study of thermally labile three-membered rink sulfones, *J. Org. Chem.*, 39, 3777, 1975.

63. Wilson, M. S. and McCloskey, J. A., Chemical ionization mass spectra of nucleosides: mechanism of ion formation and estimation of proton affinity, *J. Am. Chem. Soc.*, 97, 3436, 1975.

64. Bowen, D. V. and Field, F. H., Chemical ionization mass spectrometry: i-butane reactant gas modified by ethanolamine and ethylene diamine, *Org. Mass Spectrom.*, 9, 195, 1974.

65. Johnson, L. P., Subba Rao, S. C., and Fenselau, C., Pyridine as a reagent gas for the characterization of glucoronides by chemical ionization mass spectrometry, *Anal. Chem.*, 50, 2022, 1978.

66. Fenselau, C., Cotter, R., and Johnson, L., Mass spectral techniques for the analysis of glucoronides, *Adv. Mass Spectrom.*, 8, 1159, 1980.

67. Odiorne, T. J., Harvey, D. J., and Vouros, P., Reactions of alkyl siliconium ions under chemical ionization conditions, *J. Org. Chem.*, 38, 4274, 1973.

68. Bursey, M. M., Elwood, T. A., Hoffman, M. K., Lehman, T. A., and Tesarek, J. M., Analytical ion cyclotron resonance spectrometry. Acetylation as a chemical ionization technique, *Anal. Chem.*, 42, 1370, 1970.

69. Bursey, M. M. and Hoffman, M. K., An approach to stereochemical analysis of ion cyclotron resonance. Distinguishing *exo-* and *endo-*norborneol, *Can. J. Chem.*, 49, 3395, 1971.

70. Bursey, M. M., Kao, J. L., Henion, J. D., Parker, C. E., and Huang, T. S., Analytical ion cyclotron resonance spectrometry. Stereochemical effects in some cyclic ketones, *Anal. Chem.*, 46, 1709, 1974.

71. Hass, J. R., Bursey, M. M., and Stern, R. L., Stereospecific reactivity by ion cyclotron resonance spectrometry: optimization of reactivity differences in two isomeric esters, *Anal. Chem.*, 47, 1452, 1975.

72. Bursey, M. M., Kao, J. L., and Simonton, C. A., Analytical ion cyclotron resonance spectrometry. Stereochemical factors and a functional group interaction in some norbornyl systems, *Org. Mass Spectrom.*, 11, 149, 1976.

73. Hass, J. R., Nixon, W. B., and Bursey, M. M., Acetylation as a chemical ionization technique in medium-pressure chemical ionization mass spectrometry, *Anal. Chem.*, 49, 1071, 1977.

74. Allison, J. and Ridge, D. P., Reactions of transition metal ions with alkyl halides and alcohols in the gas phase: evidence for metal insertion and β-hydrogen atom shift, *J. Am. Chem. Soc.*, 98, 7445, 1976.

75. Allison, J. and Ridge, D. P., The gas phase chemistry of chlorotitanium ions with oxygen-containing compounds, *J. Am. Chem. Soc.*, 100, 163, 1978.

76. Allison, J. and Ridge, D. P., Reactions of atomic metal ions with alkyl halides and alcohols in the gas phase, *J. Am. Chem. Soc.*, 101, 4998, 1979.

77. Hodges, R. V. and Beauchamp, J. L., Applications of alkali ions in chemical ionization mass spectrometry, *Anal. Chem.*, 48, 825, 1976.

78. Hodges, R. V., Armentrout, P. B., and Beauchamp, J. L., Gas-phase organometallic chemistry. Reaction of Al$^+$ with alkyl halides, *Int. J. Mass Spectrom. Ion Phys.*, 29, 375, 1975.

79. Burnier, R. C., Byrd, G. D., and Freiser, B. S., Copper (I) chemical ionization mass spectrometric analysis of esters and ketones, *Anal. Chem.*, 52, 1641, 1980.

80. Cody, R. B., Burnier, R. C., Reents, W. D., Carlin, T. J., Lengel, R. K., McRery, D. A., and Freiser, B. S., Laser ionization source for ion cyclotron resonance spectroscopy. Application to atomic metal ion chemistry, *Int. J. Mass Spectrom. Ion Phys.*, 33, 37, 1980.

81. Dillard, J. G., Negative ions, in *Biochemical Applications of Mass Spectrometry*, 1st Suppl. Vol., Waller, G. R. and Dermer, O. C., Eds., John Wiley & Sons, New York, 1980, chap. 28.

82. Bowie, J. H. and Williams, B. D., Negative ion mass spectrometry of organic, organometallic, and coordination compounds, in *Mass Spectrometry*, Phys. Chem. Ser. 2, Vol. 5, MTP Rev. Sci., Maccoll, A., Ed., Butterworths, London, 1975.

83. Dougherty, R. C. and Weisenberger, C. R., Negative ion mass spectra of benzene, naphthalene, and anthracene. A new technique for obtaining relatively intense and reproducible negative ion mass spectra, *J. Am. Chem. Soc.*, 90, 6570, 1968.

84. Dougherty, R. C., Dalton, J., and Biros, F. J., Negative chemical ionization mass spectra of polycyclic chlorinated insecticides, *Org. Mass Spectrom.*, 6, 1171, 1972.

85. Hunt, D. F., Stafford, G. C., Crow, F. W., and Russell, J. W., Pulsed positive negative ion chemical ionization mass spectrometry, *Anal. Chem.*, 48, 2098, 1976.

86. Hass, J. R., Friesen, M. D., Harvan, D. J., and Parker, C. E., Determination of polychlorinated dibenzo-p-dioxin in biological samples by negative chemical ionization mass spectrometry, *Anal. Chem.*, 50, 1474, 1978.

87. Busch, K. L., Hass, J. R., and Bursey, M. M., The gas enhanced negative ion mass spectra of polychloroanisoles, *Org. Mass Spectrom.*, 13, 604, 1978.

88. Crow, F. W., Bjorseth, A., Knapp, K. T., and Bennett, R., Determination of polyhalogenated hydrocarbons by glass capillary gas chromatography-negative ion chemical ionization mass spectrometry, *Anal. Chem.*, 53, 619, 1981.

89. Hunt, D. F., McEwen, C. N., and Harvey, T. M., Positive and negative chemical ionization mass spectrometry using a Townsend discharge ion source, *Anal. Chem.*, 47, 1730, 1975.

90. Hunt, D. F. and Crow, F. W., Electron capture negative ion chemical ionization mass spectrometry, *Anal. Chem.*, 50, 1781, 1978.
91. Hunt, D. F. and Sethi, S. K., Gas-phase ion/molecule isotope-exchange reactions: methodology for counting hydrogen atoms in specific structural environments by chemical ionization mass spectrometry, *J. Am. Chem. Soc.*, 102, 6953, 1980.
92. Smit, A. L. C. and Field, F. H., Gaseous anion chemistry. Formation and reaction of OH⁻. Reactions of anions with N_2O, OH⁻ negative chemical ionization, *J. Am. Chem. Soc.*, 99, 6471, 1977.
93. Bruins, A. P., Negative ion desorption chemical ionization mass spectrometry of some underivatized glucuronides, *Biomed. Mass Spectrom.*, 8, 31, 1981.
94. Smit, A. L. C. and Field, F. H., OH⁻ negative chemical ionization mass spectrometry. Reaction of OH⁻ with methadone, L-α-acetylmethadol, and their metabolites, *Biomed. Mass Spectrom.*, 5, 572, 1978.
95. Roy, T. A., Field, F. H., Lin, Y. Y., and Smith, L. L., Hydroxyl negative chemical ionization mass spectra of steroids, *Anal. Chem.*, 51, 272, 1979.
96. Bruins, A. P., Negative ion chemical ionization mass spectrometry in the determination of components in essential oils, *Anal. Chem.*, 51, 967, 1979.
97. Houriet, R., Stahl, D., and Winkler, F. J., Negative chemical ionization of alcohols, *Environ. Health Perspect.*, 36, 63, 1980.
98. Tiernan, T. O., Chang, C., and Cheng, C. C., Formation and reactions of ions relevant to chemical ionization mass spectrometry. I. CI mass spectra of organic compounds produced by F⁻ reactions, *Environ. Health Perspect.*, 36, 47, 1980.
99. Winkler, F. J. and Stahl, D., Intramolecular ion solvation effects on gas-phase acidities and basicities. A new stereochemical probe in mass spectrometry, *J. Am. Chem. Soc.*, 101, 3685, 1979.
100. Tannenbaum, H. P., Roberts, J. D., and Dougherty, R. C., Negative chemical ionization mass spectrometry-chloride attachment spectra, *Anal. Chem.*, 47, 49, 1975.
101. Carroll, D. I., Dzidic, I., Stillwell, R. N., Horning, M. G., and Horning, E. C., New picogram detection system based on a mass spectrometer with an external ionization source at atmospheric pressure, *Anal. Chem.*, 45, 936, 1973.
102. Carroll, D. I., Dzidic, I., Stillwell, R. N., Horning, M. G., and Horning, E. C., Subpicogram detection system for gas phase analysis based upon atmospheric pressure ionization (API) mass spectrometry, *Anal. Chem.*, 46, 706, 1974.
103. Ganguly, A. K., Cappuccino, N. F., Fujiwara, H., and Bose, A. K., Convenient mass spectral technique for structural studies in oligosaccharides, *J. Chem. Soc. Chem. Commun.*, 148, 1979.
104. Dougherty, R. C. and Wander, J. D., Chloride attachment negative ion chemical ionization mass spectra of organophosphate pesticides, *Biomed. Mass Spectrom.*, 7, 401, 1980.
105. Dougherty, R. C. and Piotrowska, K., Screening by negative chemical ionization mass spectrometry for environmental contamination with toxic residues: application to human urines, *Proc. Natl. Acad. Sci. U.S.A.*, 73, 1777, 1976.
106. Bruins, A. P., Ferrer-Correia, A. J., Harrison, A. G., Jennings, K. R., and Mitchum, R. K., Negative ion chemical ionization mass spectra of some aromatic compounds using O⁻ as the reagent ion, *Adv. Mass Spectrom.*, 7, 355, 1978.
107. Goode, G. C. and Jennings, K. R., Reactions of O⁻ with some unsaturated hydrocarbons, *Adv. Mass Spectrom.*, 6, 797, 1974.
108. Dawson, J. H. J. and Jennings, K. R., Production of gas-phase radical anions by reaction of O⁻ with organic substrates, *J. Chem. Soc. Faraday Trans. II*, 72, 700, 1976.
109. Harrison, A. G. and Jennings, K. R., Reactions of O⁻ with carbonyl compounds, *J. Chem. Soc. Faraday Trans. II*, 72, 1601, 1976.
110. Jennings, K. R., Negative chemical ionization mass spectrometry, in *Mass Spectrometry*, Spec. Period. Rep. Vol. 4, Chemical Society, London, 1977, chap. 9.
111. Jennings, K. R., Investigation of selective reagent ions in chemical ionization mass spectrometry, in *High Performance Mass Spectrometry*, Gross, M. L., Ed., American Chemical Society, Washington, D. C., 1978.
112. Jennings, K. R., Chemical ionization mass spectrometry, in *Gas Phase Ion Chemistry*, Bowers, M. T., Ed., Academic Press, New York, 1979, chap. 12.
113. Dawson, J. H. J., Kaandorp, T. A. M., and Nibbering, N. M. M., A gas phase study of the ions $C_3H_2^-$ and $C_3H_3^-$ generated from the reaction of O⁻ with propyne, *Org. Mass Spectrom.*, 12, 330, 1977.
114. Dawson, J. H. J., Noest, A. J., and Nibbering, N. M. M., 1,1- and 1,3- elimination of water from the reaction complex O⁻ with 1,1,1-trideuteroacetone, *Int. J. Mass Spectrom. Ion Phys.*, 30, 189, 1979.
115. Dzidic, I., Carroll, D. I., Stillwell, R. N., and Horning, E. C., Atmospheric pressure ionization (API) mass spectrometry: formation of phenoxide ions from chlorinated aromatic compounds, *Anal. Chem.*, 47, 1308, 1975.

116. Mitchum, R. K., Althaus, J. R., Korfmacher, W. A., and Moler, G. F., Application of negative ion atmospheric pressure ionization (NIAPI) mass spectrometry for trace analysis, *Adv. Mass Spectrom.*, 8, 1415, 1980.
117. Hunt, D. F., Harvey, T. M., and Russell, J. W., Oxygen as a reagent gas for the analysis of 2,3,7,8-tetrachlorodibenzo-p-dioxin. Negative ion chemical ionization mass spectrometry, *J. Chem. Soc. Chem. Commun.*, 151, 1975.

Chapter 5

CHEMICAL IONIZATION MASS SPECTRA

I. INTRODUCTION

An astonishing number of chemical compounds has been investigated by chemical ionization methods. It is neither possible nor practical to review all of these studies. The present review will concentrate on those classes of compounds for which systematic or substantial studies have been carried out, thus permitting general conclusions to be drawn. For the most part the discussion will be limited to monofunctional compounds to illustrate the role of the functionality in determining the chemical ionization mass spectrum.

II. ALKANES (C_nH_{2n+2})

The CH_4 CI mass spectra of an extensive series of linear and branched alkanes have been reported by Field et al.[1] The spectra of the n-alkanes are characterized by abundant $[M-H]^+$ ions which account for 25 to 45% of the total additive ionization, as shown by the mass spectrum of n-decane presented as the first entry in Table 1. The $[M-H]^+$ ions originate by reaction of both CH_5^+ (Reaction 1) and $C_2H_5^+$ (Reaction 2). The CI mass spectra also show lower molecular weight alkyl ions which originate both by fragmentation of the $[M-H]^+$ ion (Reaction 3) and by nominal alkide ion abstraction reactions (Reaction 4).

$$CH_5^+ + n - C_6H_{14} \rightarrow C_6H_{13}^+ + H_2 + CH_4 \tag{1}$$

$$C_2H_5^+ + n - C_6H_{14} \rightarrow C_6H_{13}^+ + C_2H_6 \tag{2}$$

$$C_6H_{13}^+ \rightarrow C_4H_9^+ + C_2H_4 \tag{3}$$

$$\rightarrow C_3H_7^+ + C_3H_6$$

$$CH_5^+ + n - C_6H_{14} \rightarrow C_5H_{11}^+ + CH_4 + CH_4 \tag{4}$$

$$\rightarrow C_4H_9^+ + C_2H_6 + CH_4$$

$$\rightarrow C_3H_7^+ + C_3H_8 + CH_4$$

Detailed studies, using ICR techniques, have been made[2,3] of the reactions of CH_5^+ and $C_2H_5^+$ with model alkanes. These studies have shown that $C_2H_5^+$ reacts entirely by hydride ion abstraction (Reaction 2) while CH_5^+ reacts by both Reactions 1 and 4. These two reactions may be viewed either as hydride and alkide abstraction reactions or as protonation followed by fragmentation. Evidence for two forms of protonated ethane, one protonated at the C—C bond and one protonated at a C—H bond, have been presented;[4] for the higher alkanes these protonated species are thermochemically unstable with respect to fragmentation to alkyl ions. Reaction 1 could be pictured as the net result of protonation at a C—H bond while the nominal alkide abstraction reactions (Reaction 4) could equally well be pictured as the net result of protonation at the C—C bonds of the alkane. No matter how it is formed the $[M-H]^+$ ion is known to undergo fragmentation by olefin elimination to form lower alkyl ions.[5] The low energy modes of fragmentation of alkyl ions have been predicted from thermochemical arguments by Bowen and Williams.[6]

Table 1

CH₄ CI MASS SPECTRA OF ALKANES

Alkyl ion (fraction of total ionization)

Alkane	C_4	C_5	C_6	C_7	C_8	C_9	C_{10}
n-C_{10}	0.14	0.22	0.22	0.10	0.01	0.005	0.31
2-Me-5-Et-C_7	0.11	0.21	0.14	0.04	0.05	0.08	0.20
2,3,6-Me_3-C_7	0.17	0.26	0.19	0.03	0.004	0.10	0.09
2,2,4-Me_3-C_7	0.32	0.19	0.17	0.04	0.004	0.12	0.06
2,2,3,3-Me_4-C_6	0.28	0.27	0.16	0.12	0.001	0.09	0.005

Table 2

NO CI MASS SPECTRA OF ALKANES

(% of total ionization)

Alkane	$[M-H]^+$	$[M-3H]^+$	$[M-2H+NO]^+$	Alkyl ion			
				C_3	C_4	C_5	C_6
n-$C_{10}H_{22}$	80.0	2.4	1.6	2.4	2.4	1.6	9.6
n-$C_{34}H_{70}$	71.4	14.3	12.9	—	—	—	—
2,2-$Me_2C_4H_8$	86.2	—	10.3	1.7	0.9	0.9	—
2,2,4-$Me_3C_5H_{10}$	64.5	—	7.7	2.6	16.1	3.9	—
2,4,6-$Me_3C_7H_{14}$	74.1	0.7	0.1	—	3.4	10.4	11.1

The $[M-H]^+$ ion abundance for the n-alkanes is considerably greater than the ions characteristic of molecular weight (M⁺·, $[M-H]^+$) are in the electron impact mass spectrum, although in the CI spectrum the $[M-H]^+$ intensity decreases rapidly with increasing source temperature.[5] In addition, with increased branching of the alkane the $[M-H]^+$ intensity drops markedly, as the spectra in Table 1 show. Thus for 2,2,3,3-tetramethylhexane the $[M-H]^+$ ion accounts for only 0.5% of the total additive ionization. This decrease in the $[M-H]^+$ intensity with increased branching can be attributed to an increased tendency for CH_5^+ to attack the C—C bonds (Reaction 4) at the point of chain branching and to a more facile fragmentation of the $[M-H]^+$ alkyl ion, which can readily form stable secondary or tertiary alkyl ions by olefin elimination (Reaction 3).

In an attempt to obtain a greater abundance of ions characteristic of the molecular weight, Hunt and co-workers[7,8] have determined the CI mass spectra of a variety of alkanes using NO⁺ as the reagent ion. A selection of their results is presented in Table 2. The major ion observed in all spectra is the $[M-H]^+$ ion, even for branched alkanes. Thus, the NO CI mass spectra are preferred for providing molecular weight information for alkanes. A drawback to the use of NO is the formation of $[M-3H]^+$ and $[M-2H+NO]^+$ ions with many of the alkanes; ions of the same mass are observed in the NO CI mass spectra of the corresponding alkene. In the alkanes these ions probably arise by oxidation of the alkane to an alkene with subsequent reaction of the alkene to give $[M-H]^+$ and $[M+NO]^+$ ions (see the following section). The extent of oxidation is less for NO diluted with N_2 than for pure NO as the reagent gas.[7]

Alkanes are unreactive under negative ion CI conditions.

III. ALKENES AND CYCLOALKANES (C_nH_{2n})

The CH₄ CI mass spectra of a variety of alkenes have been reported by Field.[9] The spectra of monoolefins consist of two series of ions, the $[C_nH_{2n+1}]^+$ alkyl ions, consist-

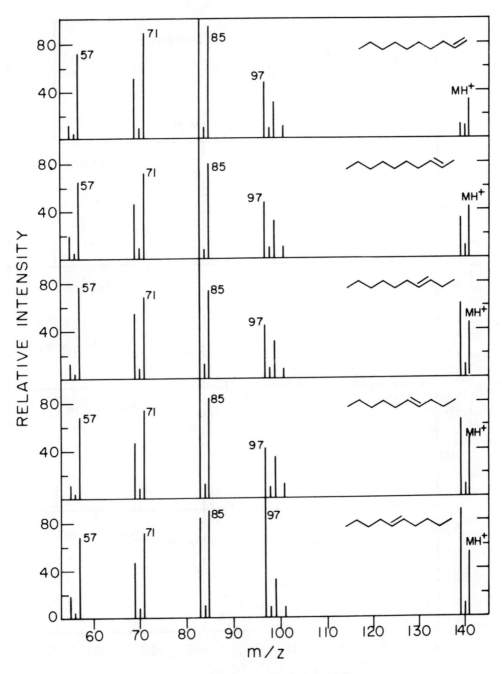

FIGURE 1.   CH₄ CI mass spectra of isomeric n-decenes.

ing of the MH⁺ ion and fragments derived therefrom, and the $[C_nH_{2n-1}]^+$ alkenyl ions, consisting of the $[M-H]^+$ ion and lower mass fragments derived therefrom. Typical spectra are shown in Figure 1 using the isomeric n-decenes as examples. In forming the $[M-H]^+$ ion allylic hydrogens apparently are most readily abstracted since the $[C_nH_{2n-1}]^+$ ions are more prominent in olefins with large numbers of allylic hydrogens. Although a full range of alkyl and alkenyl ions are formed, the intensity distribution normally peaks in the C_4 to C_7 region. The MH⁺ intensity is a maximum for C_6 olefins (15 to 20% of total ionization) and drops off with increasing olefin size to 2 to 3% of

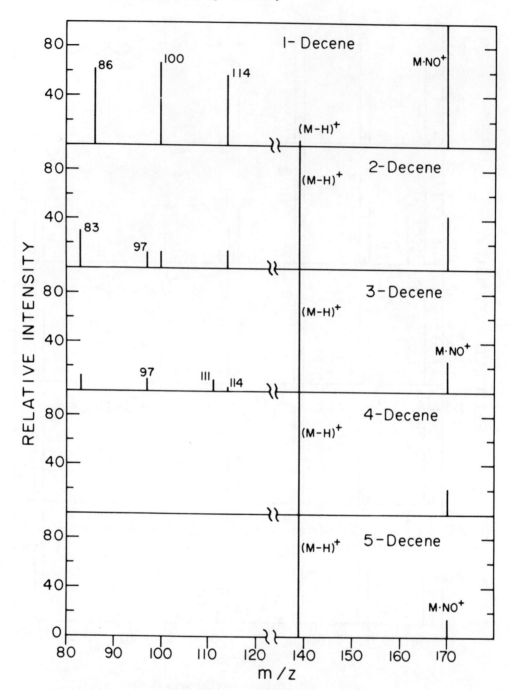

FIGURE 2.   NO CI mass spectra of isomeric n-decenes. Data from Reference 11.

total additive ionization. This intensity, combined with similar intensities for the
$[M-H]^+$ ion is sufficient to establish molecular weights. However, it should be noted
that the CH_4 CI mass spectra of alkenes give no reliable indication of the position of
the double bond or the extent and location of chain branching.

Budzikiewicz and Busker[10] have reported the i-C_4H_{10} CI mass spectra of a number
of n-octadecenes. The spectra are similar to the CH_4 CI mass spectra in showing MH^+,
$[M-H]^+$, alkyl, and alkenyl ions. In addition a prominent $[M+C_4H_9]^+$ adduct ion is

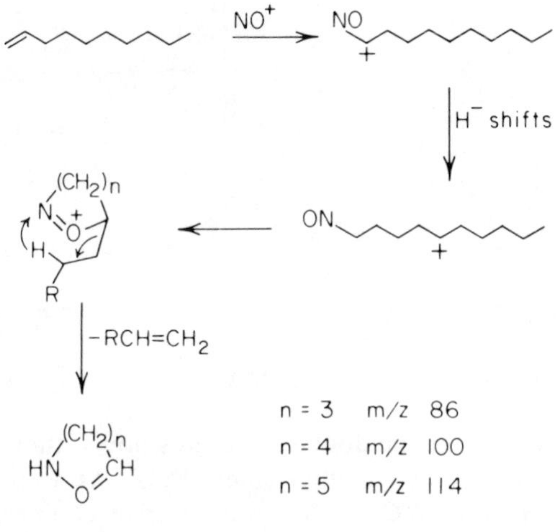

NO$^+$ → NO $+$

H$^-$ shifts

ON $+$

$-$RCH=CH$_2$

(CH$_2$)$_n$
N$+$
O
H
R

(CH$_2$)$_n$
HN CH
O

n = 3   m/z 86
n = 4   m/z 100
n = 5   m/z 114

SCHEME 1

observed. Differences in alkyl fragment ion intensities could be correlated with double bond positions for the Z-octadecenes but not, apparently, for the E-octadecenes. The Z and E isomers were distinguishable from the [M−C$_4$H$_9$]$^+$/[M−H]$^+$ ratio which was ∼ 2 for the Z isomers and <1 for the E isomers.

The NO CI mass spectra of a selection of alkenes have been reported by Hunt and Harvey.[11] The spectra reported for the n-decenes are reproduced in Figure 2. For internal olefins the spectra usually are dominated by [M−H]$^+$ ions, formed by hydride abstraction, and [M+NO]$^+$ adduct ions. If the double bond is highly substituted the ionization energy of the olefin will be lowered to the extent that charge exchange becomes exothermic and M$^{+\cdot}$ ions will be observed. For example, for 2,3-dimethyl-2-butene M$^{+\cdot}$ carries about 40% of the total additive ionization. For terminal olefins only weak [M−H]$^+$ ion signals are observed although the [M+NO]$^+$ ion signal remains strong. In addition, a series of ions C$_n$H$_{2n}$NO$^+$ (n = 3 to 6) are observed at m/z values of 7, 86, 100, and 114 as shown in Figure 2 for 1-decene. These ions are postulated[11] to arise by addition of NO$^+$ to the alkene followed by hydrogen migration and fragmentation as illustrated in Scheme 1. These C$_n$H$_{2n}$NO$^+$ ions are formed to only a minor extent for internal olefins although their abundance is enhanced at low source temperatures.[10] Budzikiewicz and Busker[10] also noted that for a number of n-alkenes the [M+NO]$^+$/[M−H]$^+$ ratio was greater (∼2) for the Z isomers than for the E isomers (<1).

In negative ion studies the OH$^-$ CI mass spectra of a few C$_8$ to C$_{10}$ alkenes have been reported by Smit and Field.[12] Although no [M−H]$^-$ ions were observed low intensity [M−3H]$^-$ ions were present in all mass spectra. The major ions observed corresponded to [M+43]$^-$ and [M+25]$^-$. The former product was postulated to arise by reaction of the initial [M−H]$^-$ product with N$_2$O to form [M−H+N$_2$O]$^-$, which by elimination of H$_2$O gives the [M+25]$^-$ product ion. The N$_2$O was present as one of the gases in the He/N$_2$O/CH$_4$ reagent mixture. In flowing afterglow studies Bohme and Young[13] have observed [M−H]$^-$ ions as the product of reaction of OH$^-$ with a number of small alkenes; their absence in the spectra reported by Smit and Field must be due to their reactivity with N$_2$O.

A problem not clearly resolved in Brønsted acid proton transfer chemical ionization, NO CI, or Brønsted base proton abstraction CI is the location of the double bond in the alkene; this problem is also not resolved by electron impact studies. In many cases, derivatization of the double bond, such as formation of a gem-diol followed by proton transfer CI[14] allows the double bond position to be determined. It obviously would be

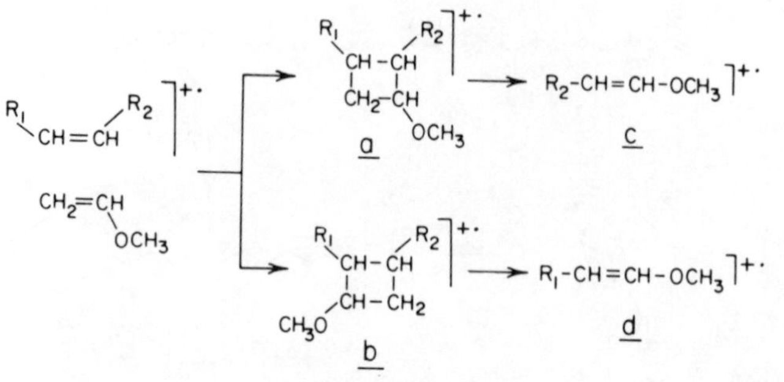

SCHEME 2

of value to be able to determine the double bond position without derivatization. This appears to be possible using vinyl methyl ether (VME) as a reagent gas, either in a mixture with CO_2[15,16] or in a N_2/CS_2/VME mixture.[17,18] The principle of the method is illustrated in Scheme 2 while typical results for the isomeric n-decenes are shown in Figure 3. Within Scheme 2, reaction of the two olefinic species, with either one bearing the charge, produces two possible cyclic complexes a and b, which on fragmentation yield, in part, the fragment ions c and d, the masses of which identify the size of R_1 and R_2, and thus the position of the double bond. The relevant fragment ions are indicated with asterisks in Figure 3. For the 1- to 3-decenes only one characteristic ion is observed, that corresponding to elimination of the smaller possible olefinic compound from the collision complex, while for 4-decene both characteristic ions are observed. Other product ions observed correspond to loss of CH_3 (m/z 183) and loss of CH_3OH (m/z 166) from the ionic collision complex. The addition of CS_2 to the reagent gas mixture simplifies the spectrum since $CS_2^{+\cdot}$, which is a low-energy charge exchange reagent ion, is formed and usually gives abundant $M^{+\cdot}$ ions on reaction with the alkene; much more extensive fragmentation is observed in the CO_2/VME mixtures since $CO_2^{+\cdot}$ is a much higher energy charge exchange reagent.

The method has been successful in locating double bonds in alkenes, unconjugated dienes, and unsaturated fatty acids. The abundance of the ions characteristic of double bond location is greater for terminal alkenes than for internal alkenes (see Figure 3) and is decreased even further for internal alkenes substituted at the double bond.[18] Thus the user of the method may encounter difficulties in the case of internal substituted alkenes.

The CH_4 CI mass spectra of a variety of cycloalkanes have been determined by Field and Munson.[19] Like their alkene isomers they show a series of alkyl, $[(C_nH_{2n+1}]^+$, ions and a series of alkenyl, $[C_nH_{2n-1}]^+$, ions, beginning with the MH^+ and $[M-H]^+$ ions, respectively. Although, in many cases, the MH^+ ion signal is low, the $[M-H]^+$ ion signal usually comprises 25 to 75% of the total additive ionization.

IV. ALKYNES, ALKADIENES, AND CYCLOALKENES (C_nH_{2n-2})

The CH_4 CI mass spectra of a few alkynes, alkadienes, and cycloalkenes have been reported by Field,[9] while the i-C_4H_{10} CI mass spectra of some octadecynes have been reported by Busker and Budzikiewicz.[20] In the CH_4 CI mass spectra two series of ions were reported, the $[C_nH_{2n-1}]^+$ ions, beginning with MH^+, and $[C_nH_{2n-3}]^+$ ions beginning with $[M-H]^+$. For the alkynes and alkadienes the former series predominated, while with the cycloalkenes the $[M-H]^+$ ion signal was particularly large. The i-C_4H_{10} CI mass spectra of the octadecynes showed the same two series of ions with the $[C_nH_{2n-1}]^+$ ions dominating. In addition $[C_nH_{2n+1}]^+$ and $[C_nH_{2n-5}]^+$ ions were observed in low abundance as well as the $[M + C_4H_9]^+$ adduct ion. The distribution of the $[C_nH_{2n-1}]^+$ ions showed some correlation with the position of unsaturation.

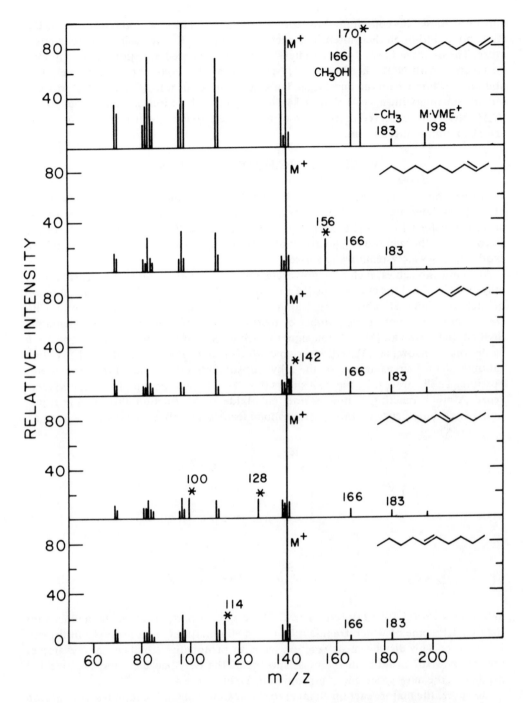

FIGURE 3. CI mass spectra of isomeric n-decenes using $N_2/CS_2/$vinyl methyl ether (VME) as reagent gas. (From Chai, R. and Harrison, A. G., *Anal. Chem.*, 53, 34, 1981. With permission.)

The NO CI mass spectra of a number of alkynes also were recorded by Busker and Budzikiewicz.[20] For dialkylacetylenes the adduct ion $[M + NO]^+$ was the base peak and no $M^{+\cdot}$ or $[M-H]^+$ ions were observed. The mass range below the molecular weight was crowded with ions of the series $[C_nH_{2n-3}]^+$, accompanied by lower-intensity $[C_nH_{2n-1}]^+$ ions and a collection of fragment ions containing NO. The relative abundances of these NO-containing ions showed some correlation with the position of the triple bond, although their intensities decreased rapidly with increasing source temperature. The n-alkynes pent-1-yne and hex-1-yne produced predominantly $[C_nH_{2n-3}]^+$ ions on reaction with NO^+; the $[M + NO]^+$, $M^{+\cdot}$, and $[M-H]^+$ ions were of extremely low abundance. For 1-heptyne and larger 1-alkynes a series of ions $[C_nH_{2n-2}NO]^+$ was observed (starting with n = 4) at m/z 84, 98, and 112. These ions are analogous to the $[C_nH_{2n}NO]^+$ ions observed on reaction of NO^+ with 1-alkenes (Scheme 1) and probably arise by a similar mechanism.

V. AROMATIC HYDROCARBONS

In their initial studies Munson and Field[21] determined the CH_4 CI mass spectra of 21 alkylbenzenes and 2 alkylnaphthalenes. Several subsequent limited studies of the proton transfer CI of alkyl benzenes have been reported,[22-25] including a study of the C_8 to C_{10} alkylbenzenes using a variety of protonating agents in which the protonation exothermicity was systematically varied.[26]

Aromatic molecules, including polycyclic aromatics, with no alkyl substituents show only MH^+ ions and the cluster ions $[M + C_2H_5]^+$ and $[M + C_3H_5]^+$. With the introduction of methyl groups H^- abstraction leading to $[M-H]^+$ also becomes possible and increases in importance as the number of methyl substituents increases. Thus pentamethylbenzene shows an $[M-H]^+$ ion signal which is 24% of the total additive ionization while toluene shows an $[M-H]^+$ ion signal which is only 3% of the total additive ionization.[21] With increasing size of the alkyl substituents other reaction modes become possible. To illustrate this, Figure 4 shows the CH_4 CI mass spectra of six C_{10} alkylbenzenes. A minor reaction channel is nominal alkide ion abstraction illustrated in Equation 5 for s-butylbenzene; this reaction mode leads primarily to benzylic-type ions at

$$(5)$$

$$(6)$$

m/z values of 91, 105, 119, etc. The $C_2H_5^+$ reacts, in part, to form the adduct ions $[M + C_2H_5]^+$ some of which fragment by olefin elimination from the alkyl substituent, Reaction 6. This displacement reaction leads to protonated ethylbenzenes; examples are the m/z 107 product in the mass spectra of the butylbenzenes and the m/z 121 product in the mass spectrum of n-propyltoluene in Figure 4.

However, the major reaction channel for both CH_5^+ and $C_2H_5^+$ continues to be proton transfer leading to MH^+. The MH^+ ion may fragment either by olefin elimination (Equation 7) or by elimination of a neutral aromatic molecule with formation of the appropriate alkyl ion (Reaction 8).

$$(7)$$

$$(8)$$

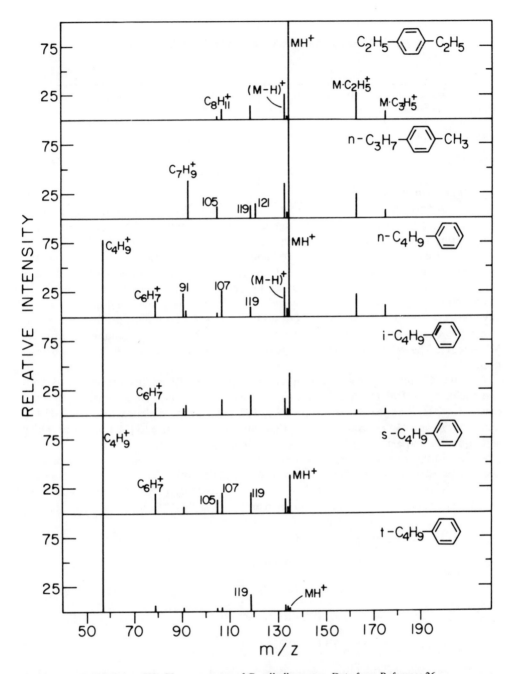

FIGURE 4.   CH$_4$ CI mass spectra of C$_{10}$ alkylbenzenes. Data from Reference 26.

Reaction 7 is responsible for the ions C$_6$H$_7^+$ (butylbenzenes), C$_7$H$_9^+$ (n-propyltoluene) and C$_8$H$_{11}^+$(diethylbenzene) in the spectra of Figure 4. From isotopic labeling studies Reaction 7 has been shown[23,25] to involve nonrandom transfer of hydrogen from several different positions of the alkyl group rather than specific migration from a single position. With increasing size of the alkyl group Reaction 8 becomes the dominant fragmentation channel and is more pronounced the more highly substituted the alkyl group is on the benzylic carbon. Thus MH$^+$ is the base peak in the CH$_4$ CI mass spectrum of n-butylbenzene but is almost entirely absent in the spectrum of t-butylbenzene.

Table 3

O⁻ CI MASS SPECTRA OF METHYLPYRIDINES

Relative intensity

Pyridine	[M−H + O]⁻ m/z 108	[M−3H + O]⁻ m/z 106	[M−CH₃ + O]⁻ m/z 94	[M−H]⁻ m/z 92	[M−2H]⁻ m/z 91	[M−3H]⁻ m/z 90
2-Methyl	100	13	13	68	95	10
3-Methyl	77	40	13	100	89	36
4-Methyl	26	15	9	100	57	9

From thermochemical considerations n-butylbenzene and i-butylbenzene would not be expected to form $C_4H_9^+$ in preference to $C_6H_7^+$ if the alkyl ions retained the primary alkyl structure;[26] it is probable that fragmentation is accompanied by hydrogen migration leading to s-$C_4H_9^+$ from n-butylbenzene and t-$C_4H_9^+$ from i-butylbenzene. As the protonation exothermicity increases, the fragmentation pattern remains the same and the MH⁺ ion intensity decreases.[26] The $C_4H_9^+$ ion (i-C_4H_{10}CI) is reported to react only very slowly with aromatic hydrocarbons;[27] the proton transfer step is endothermic.

The NO CI mass spectra of a few substituted benzenes, including alkylbenzenes, have been recorded.[28,29] The [M + NO]⁺ adduct ion is the dominant product unless the ionization energy of the aromatic is below ∼8.3 eV when charge exchange to produce M⁺˙ occurs. The NO⁺ ion does not appear to react with alkylbenzenes by H⁻ abstraction.

The OH⁻ CI mass spectra of a number of alkylbenzenes have been recorded by Smit and Field.[12] The [M−H]⁻ ion is the primary reaction product; since benzene does not react with OH⁻ it appears that it is an alkyl hydrogen which is abstracted. The [M−H]⁻ ions react to add N_2O (from the reagent gas mixture) giving rise to [M + 43]⁻ ions which subsequently, in part, fragment by loss of H_2O to give [M + 25]⁻.

Some results from a study of the reactions of O⁻ with aromatic hydrocarbons have been reported.[30-32] The products observed indicate the following sequence of reactions where R is an alkyl group and X is a halogen or hydrogen. Benzene undergoes only Reactions 10 and 11 giving peaks at m/z 76 and

$$O^- + M \;\rightarrow\; [M - H]^- + OH^\bullet \tag{9}$$

$$\rightarrow\; [M - 2H]^{-\bullet} + H_2O \tag{10}$$

$$\rightarrow\; [M - H + O]^- + H^\bullet \tag{11}$$

$$\rightarrow\; [M - R + O]^- + R^\bullet \tag{12}$$

$$\rightarrow\; [M - H + O - HX]^- + H^\bullet + HX \tag{13}$$

m/z 93, respectively, the latter accounting for about two thirds of the additive ionization. With naphthalene, Reaction 11 was by far the most important process while with anthracene both H˙ displacement (Reaction 11) and simple charge exchange to give M⁻˙ were observed. The O⁻ CI mass spectra of alkylaromatics are illustrated by the spectra of the isomeric methylpyridines given in Table 3. All the products of reactions 9 to 13 (X = H) are observed, the relative intensities depending strongly on the molecular structure. For the xylenes, [M−2H]⁻˙ is particularly prominent for the *meta* isomer; isotopic labeling has shown that one hydrogen is abstracted from each methyl group.[30]

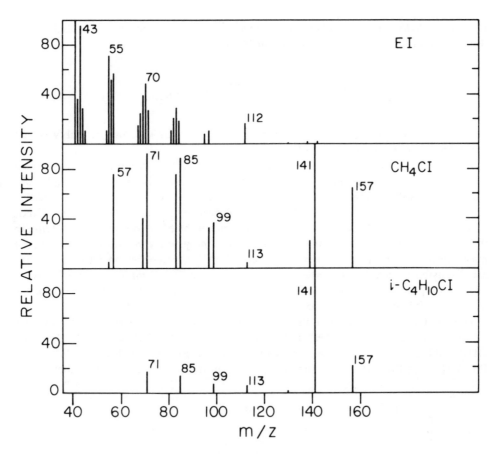

FIGURE 5. Mass spectra of n-decanol. CH₄ CI data from Reference 36, i-C₄H₁₀ CI data from Reference 33.

VI. ALCOHOLS

The t-C_4H_{10} CI mass spectra of 23 saturated, monohydroxylic alcohols have been reported by Field,[33] while the H_2 and CH_4 CI mass spectra of a series of substituted benzyl alcohols also have been reported.[34] In addition, the proton transfer CI mass spectra of isomeric C_5 and C_6 alkanols have been determined using H_3^+, N_2H^+, CO_2H^+, N_2OH^+, and HCO^+ as protonating reagents.[35] The general features of the CI mass spectra obtained using Brønsted acid reagent ions are illustrated in Figure 5 using 1-decanol as an example. For alkanols higher than C_4 the ROH_2^+ ions are not stable and form R^+ ions by loss of H_2O. Proton transfer from $C_4H_9^+$ to alkanols in i-C_4H_{10} CI is endothermic and the R^+ ion may arise by OH^- abstraction by $C_4H_9^+$. The other primary reaction channel is hydride abstraction to form $[M-H]^+$; this product normally is not observed in the CI mass spectra of tertiary alkanols indicating that the hydride abstracted originates primarily from the carbon to which the oxygen is bonded. Both the $[M-H]^+$ and R^+ ions undergo further fragmentation with the extent of fragmentation depending on the exothermicity of the initial ionization reaction;[35] this is illustrated by the lower intensity of low mass fragment ions in the i-C_4H_{10} CI mass spectrum of Figure 5. The major reaction modes of the R^+ alkyl ions are olefin eliminations to form lower mass alkyl ions while the $[M-H]^+$ ion fragments by loss of H_2O to form the alkenyl ion $[R-H_2]^+$ which may fragment further by olefin elimination. Thus the lower mass region of the CH_4 CI mass spectrum of n-decanol consists of a series of alkyl and alkenyl ions. A fragmentation reaction unique to the benzyl alcohols is the elimination of CH_2O from MH^+ (Reaction 14);

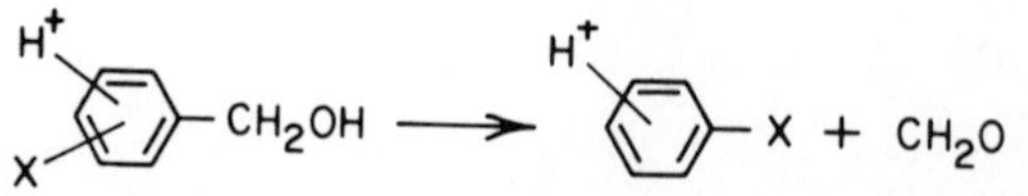

$$\overset{H^+}{\underset{X}{\diagup}}\!\!\!\diagdown\!\!\!\square\!\!\!\diagup\!\!\!-CH_2OH \longrightarrow \overset{H^+}{\diagdown}\!\!\!\square\!\!\!\diagup\!\!\!-X + CH_2O \qquad (14)$$

this reaction, which is quite pronounced in the H_2 CI mass spectra[34] probably occurs from the ring-protonated form of MH^+.

When a second functional group is present which allows intramolecular hydrogen bond formation the MH^+ ion is stabilized and, consequently, observed. Thus, in contrast to the results of Figure 5, the MH^+ ion constitutes about 14% of the total additive ionization in the CH_4 CI mass spectrum of 1,10-decanediol[37] and ~67% in the i-C_4H_{10} CI mass spectrum.[38] As a result of such stabilization the MH^+ ions are more intense in the i-C_4H_{10} CI mass spectra of the *cis* 1,3- and 1,4-cyclohexane diols than in the CI mass spectra of the *trans* isomers.[38,39] Stereochemical influences on CI mass spectra are discussed in detail in Chapter 6. The proton transfer CI mass spectra of polyhydroxyl compounds frequently show loss of more than one molecule of H_2O from the MH^+ ion.[38,40,49] (See the discussion of carbohydrates, Section XV.)

The NO CI mass spectra of a number of alcohols have been determined.[29,42] Tertiary alcohols show only an $[M-OH]^+$ ion signal. Secondary alcohols show $[M-H]^+$, $[M-OH]^+$ and $[M-2H+NO]^+$ ion signals, while primary alcohols show $[M-H]^+$, $[M-OH]^+$, $[M-3H]^+$, and $[M-2H+NO]^+$ ion signals. The primary reaction modes indicated are

$$NO^+ + H\overset{|}{\underset{|}{C}}\!-OH \rightarrow HNO + \overset{|}{C}{=}\overset{+}{OH} \qquad (15)$$

$$NO^+ + H\overset{|}{\underset{|}{C}}\!-OH \rightarrow HNO_2 + H\overset{|}{\underset{|}{C}}{}^{+} \qquad (16)$$

The $[M-2H+NO]^+$ product for secondary alcohols presumably arises from oxidation of the alcohol to a ketone followed by addition of NO^+. A similar oxidation of a primary alcohol leads to an aldehyde which reacts with NO^+ both by hydride abstraction ($[M-3H]^+$) and by NO^+ addition ($[M-2H+NO]^+$). Although these oxidation reactions eventually yield products which serve to distinguish among primary, secondary, and tertiary alcohols, the extent of oxidation probably is strongly dependent on operating conditions such as source temperature and reagent gas composition and pressure.

The reactions of OH^- with a number of alcohols have been investigated using both high-pressure CI techniques[12] and ICR techniques.[43] The dominant reaction product was found to be $[M-H]^-$, with, in a number of cases, additional formation of $[M-3H]^-$ as illustrated in Scheme 3. The ICR results suggested that the H_2 elimination reaction was collision-induced, consequently the relative intensities will be dependent on experimental conditions.

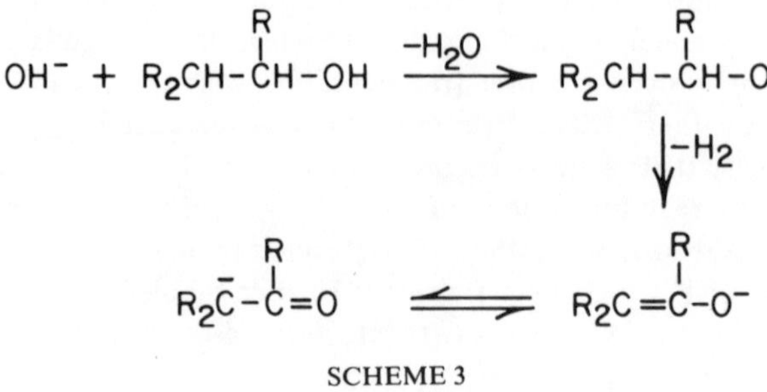

$$OH^- + R_2CH-\overset{R}{\underset{|}{C}}H-OH \xrightarrow{-H_2O} R_2CH-\overset{R}{\underset{|}{C}}H-O^-$$

$$\Big\downarrow -H_2$$

$$R_2\overset{-}{C}-\overset{R}{\underset{|}{C}}{=}O \quad \rightleftharpoons \quad R_2C{=}\overset{R}{\underset{|}{C}}-O^-$$

SCHEME 3

Table 4
O⁻ CI OF ALKANOLS

M	Relative intensity				
	$[M-H]^-$	$[M-3H]^-$	$[M-(H_2+CH_3)]^-$	$[M-(H_2+C_2H_5)]^-$	$[M-(H_2+C_3H_7)]^-$
$CH_3CH(OH)CH_3$	100	21	34	—	—
$CH_3(CH_2)_3OH$	100	37	—	—	—
$(CH_3)_2CHCH_2OH$	100	43	—	—	—
$C_2H_5CH(OH)CH_3$	100	27	23	58	—
$(CH_3)_3COH$	59	—	100	—	—
$i\text{-}C_3H_7CH(OH)CH_3$	100	13	22	—	48
$C_3H_7C(OH)(CH_3)C_2H_5$	77	—	79	100	87

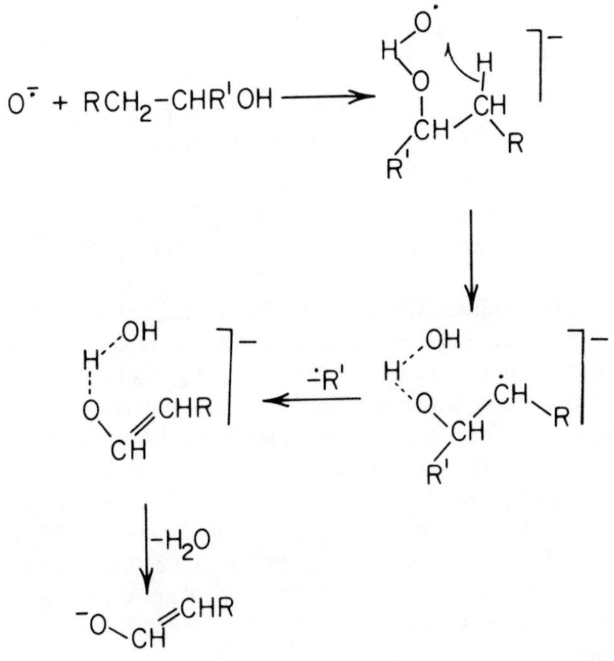

$$O^{\overline{\cdot}} + RCH_2\text{-}CHR'OH \longrightarrow$$

SCHEME 4

Much more extensive fragmentation was observed in the O⁻ CI of the few alcohols examined by Houriet et al.[43] Table 4 records the spectra reported for a selection of alcohols. At first glance the higher extent of fragmentation is surprising since O⁻ has the same proton affinity as OH⁻ (Table 6, Chapter 4). The formation of the additional fragment ions are rationalized by the mechanism shown in Scheme 4 where R′ may be a hydrogen or an alkyl group. As a consequence the O⁻ CI mass spectra provide information on the branching at the α-carbon.

VII. ETHERS

Alkyl ethers have not been studied extensively by chemical ionization techniques. The CH₄ CI mass spectrum of di-n-butyl ether is shown in Figure 6. In contrast to the analogous alcohol where MH⁺ is absent, the MH⁺ ion comprises the base peak in the spectrum of the ether. A moderate intensity [M−H]⁺ ion also is observed in the molecular weight region. The major fragment ion is the alkyl ion formed by alcohol elimi-

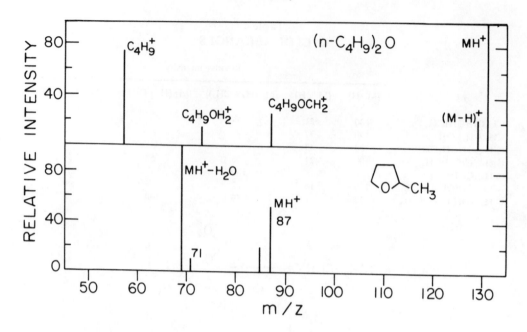

FIGURE 6.   CH₄ CI mass spectra of di-n-butyl ether and 2-methyltetrahydrofuran.

nation from MH⁺; olefin elimination to form the protonated alcohol also occurs. There also is a minor reaction channel involving alkide ion abstraction to form $C_4H_9OCH_2^+$.

The i-C₄H₁₀ CI of a number of cyclic ethers has been studied;[44] abundant MH⁺ ions are observed with the major fragmentation channel involving loss of H_2O from MH⁺. This fragmentation mode also is observed as the main fragmentation channel of MH⁺ in a study of proton transfer to 2-methyloxetane, 3-methyloxetane, and tetrahydrofuran using a variety of protonating agents.[45] The CH₄ CI mass spectrum of 2-methyltetrahydrofuran is shown in Figure 6. The proton transfer CI mass spectrum of phenyl propyl ether shows MH⁺ as the base peak with the major fragmentation channel involving propene elimination forming protonated phenol; isotopic labeling studies[46] show that the migrating hydrogen comes from each position of the propyl group in a nonrandom fashion.

The NO CI mass spectrum of di-n-amyl ether shows [M−H]⁺ as the sole product ion;[29] this probably is characteristic of ethers.

VIII. ALDEHYDES AND KETONES

The proton transfer chemical ionization of aldehydes and ketones has not been systematically studied. Protonation of simple C₃ and C₄ aldehydes and ketones by a variety of protonating agents ranging from H_3^+ to HCO⁺ has been studied.[45,47] The CH₄ CI mass spectrum of n-heptanal also has been reported.[48] The CH₄ CI mass spectra of 4-heptanone and n-heptanal are shown in Figure 7. For the ketone, MH⁺ is the dominant peak with only minor fragmentation by elimination of H_2O. By contrast, for the aldehyde extensive elimination of H_2O from MH⁺ has occurred. The mechanism of H_2O elimination from the MH⁺ ions of aldehydes has been investigated by deuterium labeling[48,49] and shown to be quite complex; the reagent proton is not always lost in the elimination process.

The NO CI mass spectra of a relatively few aldehydes and ketones have been reported.[29,49] The spectra of aldehydes normally show [M−H]⁺ ions and either [M + NO]⁺

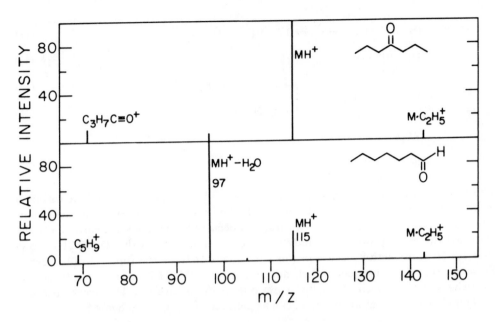

FIGURE 7.   CH₄ CI mass spectra of 4-heptanone and n-heptaldehyde.

Table 5
O⁻ CI OF KETONES

	Relative intensity					
RCOR′	[M−H]⁻	[M−H₂]⁻̇	RCOO⁻	R′COO⁻	[M−H₂−R]⁻	[M−H₂−R′]⁻
CH₃COCH₃	100	84	2	a	55	a
C₂H₅COCH₃	100	20	3	3	11	6
n-C₃H₇COCH₃	100	31	5	4	7	15
i-C₃H₇COCH₃	100	4	5	5	18	—
t-C₄H₉COCH₃	100	—	—	—	27	—
C₂H₅COC₂H₅	100	24	10	a	27	a
n-C₃H₇COn-C₃H₇	100	28	34	a	—	—
i-C₃H₇COi-C₃H₇	100	—	48	a	—	—

ᵃ   R = R′.

or M⁺· ions depending on the ionization energy of the aldehyde. Ketones do not show [M−H]⁺ ions but only [M + NO]⁺ or M⁺· ions depending on the ionization energy of the ketone. In both cases charge exchange is not important unless the ionization energy of the sample is below ~8.5 to 8.8 eV.

The OH⁻ CI mass spectra of a few ketones have been reported.[12] The major product ion is [M−H]⁻ with a minor yield of [M−3H]⁻.

The reactions of O⁻ with a variety of simple ketones have been investigated by ICR techniques[50] while the O⁻ CI mass spectra of some C₅ ketones have also been reported.[31] The same reactions occur in both systems, however the relative yields are rather different. The relative product ion intensities observed in the ICR study for a number of ketones are recorded in Table 5. The products observed can be rationalized by the following reaction scheme:

$$O^{\overline{\cdot}} + RCOR' \rightarrow (M-H)^{\cdot} + OH^{-} \xrightarrow{RCOR'} [M-H]^{-} + H_2O \tag{17}$$

$$\rightarrow [M-H]^{-} + OH^{\cdot} \tag{18}$$

$$\rightarrow H_2O + [M-H_2]^{\overline{\cdot}} \longrightarrow \begin{cases} [M-H_2-R]^{-} + {\cdot}R' & (19) \\ [M-H_2-R']^{-} + {\cdot}R & (20) \end{cases}$$

$$\rightarrow RCOO^{-} + {\cdot}R' \tag{21}$$

$$\rightarrow R'COO^{-} + {\cdot}R \tag{22}$$

The formation of the carboxylate anions serves to locate the position of the carbonyl group. Of note is the fact that di-i-propyl ketone shows no $[M-H_2]^{\overline{\cdot}}$ fragment. This was taken to indicate that the $H_2^{\overline{\cdot}}$ abstraction occurred from the same carbon α to the carbonyl function; however, a study[51] of the reaction of $O^{\overline{\cdot}}$ with CD_3COCH_3 shows that 1,3-$H_2^{\overline{\cdot}}$ abstraction occurs in this case about 56% of the time. It still appears, however, that the subsequent alkyl radical loss from $[M-H_2]^{\overline{\cdot}}$ occurs from those species where $H_2^{\overline{\cdot}}$ has been abstracted from one of the α carbons. The overall process, using n-$C_3H_7COCH_3$ as an example, can be pictured as shown in Scheme 5.

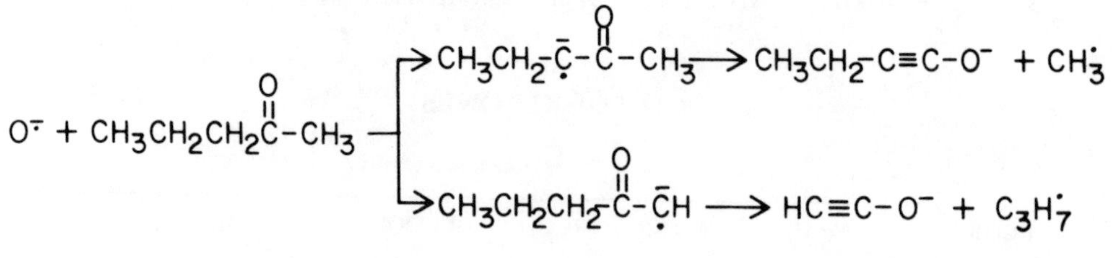

SCHEME 5

IX. HALOGENATED COMPOUNDS

There have been no systematic studies of the chemical ionization of simple haloalkanes. In proton transfer CI hydrogen halide (HX) should be readily lost from the initially formed protonated species RXH^+; in addition further fragmentation of R^+ will occur if the protonation step is sufficiently exothermic. In line with these predictions, in the H_2 CI mass spectra of simple alkyl chlorides no RXH^+ ion is observed, except for methyl chloride.[52] Similarly, the reaction of CH_5^+ with the propyl chlorides leads to formation of $C_3H_7^+$ rather than the protonated chloride.[53] The CH_4 CI mass spectra of the cyclohexyl halides show only $[MH^+-HX]$ (X = Cl, Br, I)[54] while the H_2 and CH_4 CI mass spectra of the benzyl halides show $[MH^+-HX]$ as the only significant ion signal.[55,56]

When the halogen is attached to an aromatic nucleus a much more complex chemistry results. The proton transfer CI of halobenzene derivatives has been studied in considerable detail;[55-61] valuable mechanistic information also has been derived from radiation chemical[62,63] and ICR[64] studies. Figure 8 shows the H_2 CI mass spectra of the halobenzenes. Appreciable MH^+ ion signals are observed, except for iodobenzene, despite the strongly exothermic protonation reaction. In addition, charge exchange from H_3^+ to form $M^{+\cdot}$ occurs; this is a common reaction mode of H_3^+. The major fragmentation reactions of MH^+ involve loss of a neutral hydrogen halide to form the phenyl cation or loss of a halogen atom to form the benzene molecular ion. This latter reaction is one of the few examples of formation of an odd-electron product in proton

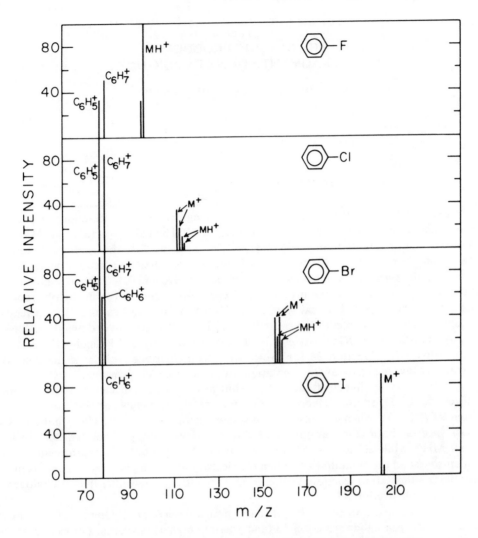

FIGURE 8. H₂ CI mass spectra of halobenzenes. Data from References 55 and 56.

transfer chemical ionization. The phenyl cation adds to H_2 to form the protonated benzene ion. The following reaction scheme is indicated where AH is the reagent gas, AH_2^+ the gaseous Brønsted acid, X is a halogen, and R is a substituent on the phenyl ring

$$AH_2^+ + RC_6H_4X \xrightarrow{-AH} RC_6H_4X \cdot H^{+*} \qquad (23)$$

$$RC_6H_4X \cdot H^{+*} \xrightarrow{M} RC_6H_4X \cdot H^{+} \qquad (24)$$

$$RC_6H_4X \cdot H^{+*} \longrightarrow RC_6H_5^{+ \cdot} + X^{\cdot} \qquad (25)$$

$$RC_6H_4X \cdot H^{+*} \longrightarrow RC_6H_4^+ + HX \qquad (26)$$

$$RC_6H_4^+ + AH \longrightarrow RC_6H_4A \cdot H^+ \qquad (27)$$

Table 6
ΔH_f° OF IONIC PRODUCTS IN
FRAGMENTATION OF $C_6H_5X \cdot H^{+a}$

X	ΔH_f° ($C_6H_6^{+\cdot}$ + X$\cdot$)	ΔH_f° ($C_6H_5^+$ + HX)
F	252	201
Cl	262	244
Br	260	257
I	259	281

ᵃ   Data in kcal mol⁻¹.

In the H_2 CI system the competition between fragmentation by Reactions 25 and 26 is determined primarily by the relative energetics.[56,58] The $C_6H_6^{+\cdot}$ ion (ΔH_f° = 233 kcal mol⁻¹ [65]) is more stable than $C_6H_5^+$ (ΔH_f° = 266 kcal mol⁻¹)[66] and tends to drive the reaction to form $C_6H_6^{+\cdot}$ + X$\cdot$, with the balance between the two reactions being determined by the relative thermochemical stabilities of HX and X$\cdot$. Table 6 summarizes the heats of formation of the products of Reactions 25 and 26 for the halobenzenes. Because of the thermochemical stability of HF (ΔH_f° = −65 kcal mol⁻¹),[65] Reaction 26 is the preferred fragmentation route for $C_6H_5F \cdot H^+$. As one moves down the halogen family the stability of XH relative to X$\cdot$ decreases markedly and Reaction 25 becomes competitive with Reaction 26 for protonated bromobenzene while it is the thermochemically favored route of fragmentation for protonated iodobenzene.

An electron-releasing substituent, R, stabilizes the $RC_6H_5^{+\cdot}$ species more than it stabilizes the $RC_6H_4^+$ species. Hence there is an increased tendency for elimination of X$\cdot$ from $RC_6H_4X \cdot H^+$ when electron-releasing substituents are present. Thus, in the H_2 CI mass spectra, both chloroanisoles and chloroanilines show predominant loss of Cl$\cdot$ from MH$^+$, while Br loss is the only significant fragmentation for protonated bromoanisoles and bromoanilines.[58] When the electron-attracting nitro substituent is present the fragmentation reactions center around the nitro group rather than the halogen[69] as discussed in Section XI.

Two further features of the H_2 CI mass spectra deserve note. One is the sequential loss of hydrogen halide from dihalo (and possibly higher) benzenes. For example, the $ClC_6H_5^+$ ion formed from dichlorobenzene or chlorofluorobenzene does not form a stable adduct on reaction with H_2; rather the adduct eliminates HCl to form $C_6H_5^+$, which reacts with H_2 to form $C_6H_7^+$. This sequence of reactions adds to the complexity of the H_2 CI mass spectra of polyhalobenzenes and also renders the relative intensities sensitive to reagent gas pressure. Finally, the $RC_6H_4^+$ cations frequently show different reactivities toward H_2. Thus the [MH⁺−HCl] ion from 2-chlorobiphenyl is essentially unreactive, that from 3-chlorobiphenyl is very reactive, and that from 4-chlorobiphenyl is moderately reactive towards H_2. As a result the three isomers can be distinguished by their H_2 CI mass spectra.[68] Similar differences are observed for the dichlorobiphenyls but not for the more highly chlorinated species.[68]

The CH_4 CI mass spectra of the halobenzenes and halotoluenes show[56,58] essentially the same features as the H_2 CI mass spectra, however, the extent of fragmentation is much less since the protonation exothermicity is lower. The similarity in behavior includes addition of the phenyl cations to CH_4 (Reaction 27). Features specific to the CH_4 CI are the formation of the cluster ions $[M + C_2H_5]^+$ and $[M + C_3H_5]^+$ and the displacement of an iodine atom by $C_2H_5^+$

$$C_2H_5^+ + RC_6H_4I \rightarrow C_2H_5C_6H_4R^{+\cdot} + I^\cdot \qquad (28)$$

However, significant differences in the character of the CH_4 and H_2 CI mass spectra are observed for halobenzenes bearing electron-releasing substituents. Table 7 presents the CH_4 CI mass spectra of the isomeric fluoro-, chloro-, and bromo-anilines.[59] The features of note are the almost complete absence of loss of HX from MH[+] for *ortho* substituted compounds and the significant enhancement of this process for *meta* substituted compounds. This is particularly noticeable for the chlorobenzene derivatives where [MH[+]−HCl] and its adduct to CH_4 are observed only when the chlorine is *meta* to the electron-releasing substituent. This behavior has been observed for a large variety of substituted chlorobenzenes and formation of [MH[+]−HCl] occurs in the CH_4 CI of the *meta* compounds even when formation of [MH[+]−Cl] is thermochemically more favorable.[59] These specific substituent effects have been attributed[59] to ion-local bond dipole interactions which influence the extent of protonation at the halogen, although it is clear that there are a variety of subtle structural and energetics effects which also play a role.[61] The CI mass spectra obtained using CO_2H^+ as reagent ion are similar to the CH_4 CI mass spectra;[60] the proton affinity of CO_2 is the same as $PA(CH_4)$. Using N_2H^+ as reagent ion the CI mass spectra are intermediate between the H_2 CI and CH_4 CI mass spectra,[60] as expected from the intermediate proton affinity of N_2. In a number of systems positional isomers appear identifiable from the N_2H^+ CI mass spectra.

There have been numerous investigations of the proton transfer CI mass spectra of polyhalogenated compounds largely as a result of their role as significant environmental pollutants. Biros et al.[69] and McKinney et al.[70] have reported the CH_4 CI mass spectra of a series of polycyclic nonaromatic chlorinated insecticides and metabolites. Not surprisingly, the MH[+] ion intensities were very small, the base peak normally being [MH[+]−HCl] except for those compounds containing a free hydroxyl group where [MH[+]−H_2O] was the base peak. Further loss of HCl from [MH[+]−HCl] was observed as was a retro-Diels-Alder reaction. It was suggested that the relative importance of the latter reaction might be useful in distinguishing between isomers, viz., aldrin vs. isodrin, α-chlordane vs. γ-chlordane. The i-C_4H_{10} CI of this class of compounds showed a similar behavior but with less extensive fragmentation.[70] Similarly in the CH_4 CI mass spectra of the isomeric hexachlorocyclohexanes no MH[+] ions were observed, the major ion signals corresponding to [MH[+]−HCl] and [MH[+]−2HCl] with the latter the base peak.[71] The major fragment ions in the i-C_4H_{10} CI mass spectra were reported to be [M[+·]−HCl] rather than [MH[+]−HCl].[71] Proton transfer from $C_4H_9^+$ to $C_6H_6Cl_6$ undoubtedly is endothermic and the ionization observed may reflect charge exchange from minor ions in the CI plasma rather than reaction of the major plasma ion $C_4H_9^+$. The authors noted that larger samples were necessary to obtain i-C_4H_{10} CI mass spectra, consistent with an ionization of low probability because of the lower concentration of reactive ions.

Despite the presence of the basic aromatic system the CH_4 CI mass spectra of compounds of the type $(ClC_6H_4)_2CHR$ (R = CCl_3, $CHCl_2$, CH_2Cl, CH_3) showed no MH[+] ion signals; rather complete fragmentation occurred forming [MH[+]−HCl], [MH[+]−RH], and [MH[+]−ClC_6H_5], with the latter comprising the base peak.[70] The i-C_4H_{10} CI mass spectra of a number of these compounds have been reported[72] and show a high abundance for the [MH[+]−HCl] ion. Curiously, the [M[+·]−HCl] ion was reported to be the base peak for p,p'-DDT whereas [MH[+]−HCl] was the base peak for the remaining compounds including o,p'-DDT. The related olefinic compounds $(ClC_6H_4)_2C=R$ (R = CCl_2, CHCl, CH_2) show more pronounced MH[+] ion signals in their CH_4 CI mass spectra with the major fragmentation reactions being elimination of HCl and C_6H_5Cl from MH[+].[70]

The CH_4 CI mass spectra of the polybromobenzenes and the polychlorobenzenes are dominated by MH[+] ions and the adduct ions [M + C_2H_5][+] and [M + C_3H_5][+].[73] Similar results are obtained for the polychlorobiphenyls.[68,74] In general it appears that pro-

Table 7

CH$_4$ CI MASS SPECTRA OF HALOANILINES

Halogen				Relative intensity				
	M·C$_3$H$_5^+$	M·C$_2$H$_5^+$	MH$^+$	M$^{..}$	[M·C$_2$H$_5^+$–X]	[MH$^+$–X]	[MH$^+$–HX]	[MH$^+$–HX + CH$_4$]
o-F	4.3	11.2	100	8.9	—	—	3.1	—
m-F	9.6	33.6	100	17.2	—	0.9	24.6	22.2
p-F	7.9	27.5	100	28.9	—	5.8	6.7	10.6
o-Cl	7.8	18.6	100	9.3	—	—	—	—
m-Cl	7.1	20.3	100	15.0	—	—	18.6	17.9
p-Cl	7.1	19.6	100	9.7	—	—	1.8	1.8
o-Br	4.7	6.9	100	11.3	18.2	24.3	—	—
m-Br	8.1	4.6	100	16.2	33.2	62.5	10.9	15.9
p-Br	1.6	2.4	58.2	28.2	26.9	100	0.2	—

Table 8
ELECTRON CAPTURE CI MASS
SPECTRA OF TRICHLOROANISOLES

Trichloroanisole	Relative intensity		
	[M−H]⁻	[M−CH₃]⁻	[M−HCl]⁻
2,3,4-	30	25	100
2,3,5-	40	30	100
2,3,6-	—	—	100
2,4,5-	55	20	100
2,4,6	—	—	100
3,4,5-	100	70	—

ton transfer CI offers little advantage over electron impact in the identification of polyhalogenated compounds.

The electron capture CI mass spectra (methane moderating gas) of a series of bromo- and chlorobenzenes have been reported.[73] For the lower bromobenzenes Br⁻ was the dominant production; however, for hexabromobenzene some M⁻ (∼30% of additive ionization) was observed. Similarly, for the lower chlorobenzenes Cl⁻ was the dominant ion observed, with M⁻ being dominant (>90% of additive ionization) for C_6HCl_5 and C_6Cl_6. When oxygen was added to the methane moderating gas no significant differences were observed in the spectra of the bromobenzenes. For the chlorobenzenes there was an increased abundance of [M−Cl]⁻ for the lower members of the series and formation of [M−Cl+O]⁻ for the higher members. The displacement of Cl by O⁻ was reported earlier[75] in an atmospheric pressure ionization study of the chemical ionization of polychlorinated compounds using air or N_2 (containing 0.5 ppm O_2) as carrier gas. The phenoxide ions are formed by either or both of the reactions

$$M^{\overline{\cdot}} + O_2 \rightarrow [M - Cl + O]^- + \dot{O}Cl \tag{29}$$

$$O_2^{\overline{\cdot}} + M \rightarrow [M - Cl + O]^- + \dot{O}Cl \tag{30}$$

Busch et al.[76] have reported the electron capture CI mass spectra (methane moderating gas) of a selection of tri- and tetrachloroanisoles. The spectra of the trichloroanisoles are presented in Table 8. The relative intensities of the [M−H]⁻, [M−CH₃]⁻, and [M−HCl]⁻ ions observed are dependent on the isomer. Elimination of H from the molecule anion occurs only when there is an H *ortho* to OCH₃; this suggests that it is an *ortho* H which is eliminated. Elimination of HCl occurs only when there is a Cl *ortho* to the OCH₃; isotopic labeling showed that the hydrogen of the HCl originated from the OCH₃ group. Similar reactions were observed for the tetrachloroanisoles. Addition of oxygen to the methane gas resulted in several new products, the major one being an [M−55]⁻ ion postulated to arise as illustrated in Scheme 6 by reaction of [M−HCl]⁻ with O_2.

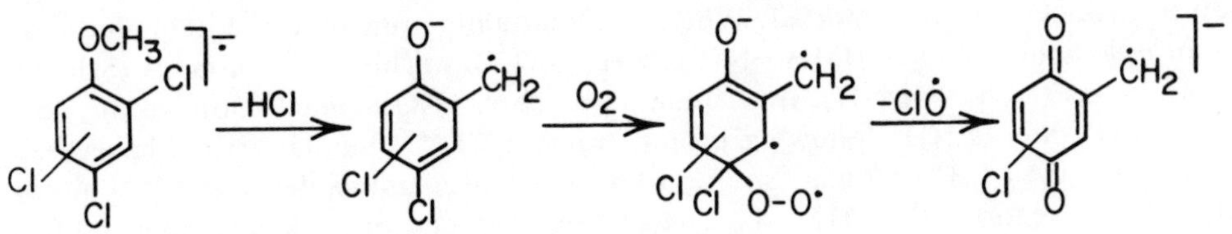

SCHEME 6

In similar studies the electron capture CI mass spectra of chlorinated diphenyl ethers[77] and chlorinated 2-phenoxyphenols[78] have been reported. In none of these studies was any mention made of the intensity of the Cl^- product; formation of significant amounts of this noninformative product could considerably reduce the sensitivity of the CI process.

The O_2 negative ion CI mass spectrum of 2,3,7,8-tetrachlorodibenzo-p-dioxin (TCDD) shows[79] formation of the molecular anion and phenoxide anions arising by replacement of both H and Cl by O^-. However, by far the most intense product ion is an *o*-quinone ion at m/z 176 postulated to arise as shown in Scheme 7, although it may also arise by reaction of O_2^- with neutral M. Formation of this product has been used to monitor TCDD in environmental samples since it appears to be clear of other interfering polychlorinated compounds.[80,81]

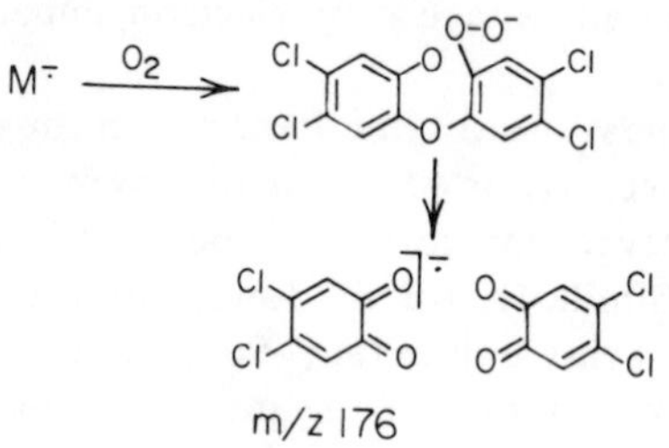

m/z 176

SCHEME 7

Dougherty and colleagues[72,82] have determined the negative ion CI mass spectra of a number of polycyclic chlorinated and aromatic chlorinated pesticides. Using either CH_4 or i-C_4H_{10} as moderating gases the major feature in all mass spectra was the group of isotopic peaks due to Cl^- attachment to the neutral molecule, together with low intensity M^- ions for the polycyclic compounds. It appears that the near-thermal electrons react rapidly by dissociative attachment to give Cl^-. The main features of the CI mass spectra observed result from reaction of this product ion. The use of CH_2Cl_2 as reagent gas, where Cl^- should be formed more directly, resulted in very similar CI mass spectra.[83]

X. AMINES

The CI mass spectra of simple amines have not been studied extensively. In terms of the discussion of Section II.A., Chapter 4, loss of NH_3 from $RNH_2 \cdot H^+$ is not a facile process and appreciable MH^+ ion intensities should be seen in proton transfer CI mass spectra. In line with these predictions the CH_4 CI mass spectrum of cyclohexylamine shows MH^+ (100%) $[M-H]^+$ (70%) and $[MH^+-NH_3]$ (59%).[54] Similarly, in the CH_4 CI mass spectra the $[MH^+-NH_3]/MH^+$ ion intensities were reported to be 0.04 for n-butylamine, 0.07 for s-butylamine and 2.3 for t-butylamine;[84] the high stability of the t-butyl cation overcomes the poor leaving tendency of NH_3. The $[M-H]^+$ ion intensities were not reported. When a second amine function is suitably located, as in α,ω-diaminobutane ($[MH^+-NH_3]/MH^+$ = 2.3), anchimeric assistance (Scheme 8) promotes the loss of NH_3. In α,ω-amino alcohols H_2O is much more readily lost from MH^+ than is NH_3 following proton transfer CI.[84,85] Similar results have been obtained in the i-C_4H_{10} CI mass spectra of a variety of β-amino alcohols.[86-88] Whitney et al.[89] have reported the CH_4 and i-C_4H_{10} CI mass spectra of trimethylamine and five polytertiary amines. In the i-C_4H_{10} CI mass spectra the dominant ion was MH^+ (60 to 93% of additive ionization). In the CH_4 CI mass spectrum of $(CH_3)_3N$ MH^+ (45%)

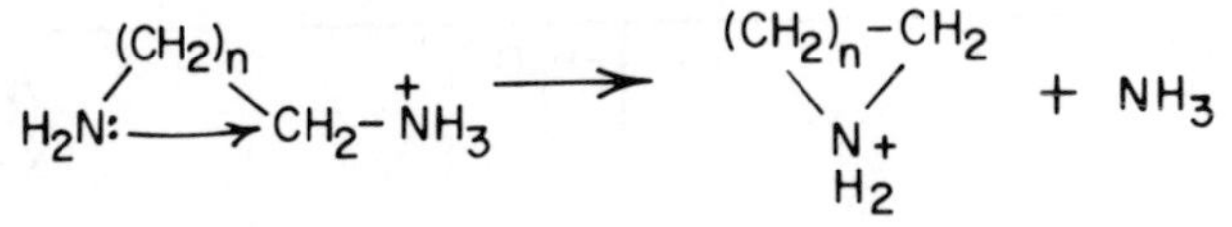

$$H_2N: \longrightarrow CH_2 - \overset{+}{N}H_3 \xrightarrow{\quad} \overset{(CH_2)_n - CH_2}{\underset{\underset{H_2}{N+}}{\diagdown \diagup}} + NH_3$$

with $(CH_2)_n$ shown above the left structure.

SCHEME 8

and $[M-H]^+$ were both abundant. For the polyamines the $[M-H]^+$ ion varied from 25 to 44% of the additive ionization and was usually slightly larger than the MH^+ ion signal. The major fragmentation reactions of MH^+ involved elimination of neutral amines and it appeared that anchimeric assistance similar to Scheme 8 was involved.

XI. NITRO COMPOUNDS

It has been shown[90,91] that nitroaromatic compounds can undergo reduction to the corresponding amine under proton transfer CI conditions using H_2, CH_4, i-C_4H_{10}, or NH_3 as reagent gases. Protonation of the amine gives an MH^+ ion which is isobaric with the $[MH^+-NO]$ ion which can be formed by fragmentation of the protonated nitro compound. This reduction reaction is favored by high source temperatures and the presence of water in the ion source.[67]

The H_2 and CH_4 CI mass spectra of a selection of substituted nitrobenzenes have been reported[67] where precautions were taken to avoid this reduction process. The greater extent of fragmentation in the H_2 CI mass spectra made these spectra more useful in distinguishing among isomers. The H_2 CI mass spectra of the three nitroanilines are shown in Figure 9. When there is an *ortho* substituent bearing a hydrogen a significant peak is observed corresponding to $[MH^+-H_2O]$ (Scheme 9).

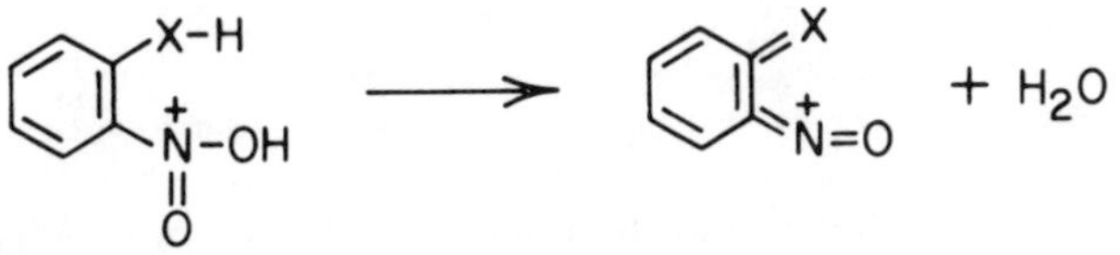

SCHEME 9

Other fragmentation modes involve loss of OH, NO, HNO, NO_2, and HNO_2 from the MH^+ ion. Loss of OH is particularly prevalent when an electron-releasing substituent is *ortho* or *para* to the nitro group, while loss of fragments containing NO or NO_2 is particularly pronounced when an electron-releasing substituent is *meta* to the nitro group. Thus, isomeric nitroarenes containing electron-releasing substituents can be identified; identification of isomers containing electron-attracting substituents is much more problematical. The extent of fragmentation of MH^+ in the CH_4 CI mass spectra was very small unless elimination of H_2O as in Scheme 9 was possible.

The CH_4 and i-C_4H_{10} CI mass spectra of a series of trinitroaromatic compounds also have been reported.[92] The base peak normally is MH^+ with minor fragmentation by loss of OH, H_2O, and, apparently, NO from MH^+. It has been shown[93] that reduction to an amine also occurs for the trinitroaromatics and the $[MH^+-30]$ ion intensities reported in the earlier study may represent protonation of the amine rather than loss of NO from MH^+.

The CH_4 CI mass spectra of various monitro-, gem-dinitro-, and 1,1,1-trinitroalkanes, halogenonitromethanes, tetranitromethane, and phenylnitromethanes have been reported.[94] In contrast to the EI mass spectra which give no molecular ions, the CI mass spectra give abundant MH^+ ions except for the phenylnitromethanes which

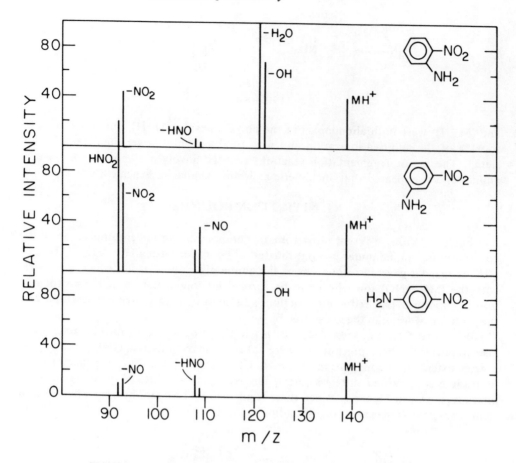

FIGURE 9. H_2 CI mass spectra of nitroanilines. Data from Reference 67.

eliminate HNO_2 forming stable benzylic ions. Fragmentation of MH^+ was by loss of H_2O, HNO, and HNO_2; the NO_2^+ ion was abundant in the spectra of the trinitroalkanes.

XII. CARBOXYLIC ACIDS AND ESTERS

The CH_4 CI mass spectra of carboxylic acids normally show appreciable MH^+ ion signals with the dominant fragmentation reactions corresponding to loss of H_2O from MH^+.[34,95,96] Benzoic acids also show loss of CO_2 from MH^+ resulting in a protonated benzene ion.[34] The elimination of H_2O is enhanced substantially by interaction with a second carboxyl group. Thus, dodecanoic acid shows an $[MH^+-H_2O]/MH^+$ ratio of 0.7 while dodecan-1,10-dioic acid shows a ratio of 50.[95] Similarly, fumaric acid, where the two carboxyl groups cannot interact, shows a $[MH^+-H_2O]/MH^+$ ratio of 0.2 compared to a ratio of 10 for maleic acid.[96] In the same vein H_2O loss is much more pronounced from the MH^+ ion of phthalic acid than from the MH^+ ion of isophthalic acid.[96] In protonated benzoic acids loss of H_2O is more pronounced in those acids bearing an *ortho* substituent bearing a lone pair of electrons, e.g., OH, NH_2, F, Cl.[34]

The thermochemically favored site of protonation of a carboxylic acid is the carbonyl oxygen.[97] Loss of H_2O from this species requires (Reaction 31) a 1,3-H shift, which is symmetry-forbidden and has a large energy barrier.[98]

$$\underset{\substack{\text{OH} \\ | \\ \text{R}-\text{C}\overset{+}{\cdots}\overset{+}{\text{OH}}}}{} \rightarrow \underset{\substack{\text{O} \\ \| \\ \text{R}-\text{C}-\overset{+}{\text{OH}}_2}}{} \rightarrow \text{R}-\text{C}\equiv\text{O}^+ + \text{H}_2\text{O} \qquad (31)$$

A second interacting carboxyl group obviates the necessity for such a 1,3-H shift and provides a lower energy route to fragmentation (Reaction 32). In addition, the product likely is stabilized by formation of the protonated anhydride. Similarly, in protonated benzoic acids an *ortho* substituent capable of accepting a proton acts as an intramolecular catalyst for the hydrogen shift necessary for H_2O elimination (Reaction 33).

$$(32)$$

$$(33)$$

In the CH_4 CI of methyl esters the dominant fragmentation reaction of MH^+ becomes elimination of CH_3OH and this fragmentation reaction is similarly catalyzed by a second carboxyl group.[95,96] For a number of α,β-unsaturated esters an alternative mode of fragmentation of MH^+ was observed[99] to be ketene elimination as illustrated in Reaction 34 for methyl cinnamate; this reaction presumably follows protonation at the double bond.

$$(34)$$

The H_2 and CH_4 CI mass spectra of a series of methyl n-alkanoates up to methyl stearate have been reported.[100] With increasing size of the alkyl chain, reactions characteristic of an alkane, i.e., H^- abstraction, are observed as well as reactions characteristic of the carboxyl function. In addition to loss of CH_3OH from MH^+, loss of CH_3OH from $[M-H]^+$ also is observed. Figure 10 shows the dependence of ion abundances on chain length in the H_2 CI. The CH_4 CI system shows a similar behavior except that the abundance of fragment ions is lower and the $[M^.+C_2H_5]^+$ and $[M+C_3H_5]^+$ cluster ions are observed.[100]

The CI mass spectra of a number of dicarboxylic acids and their methyl esters have been determined using the reagent ions M_3O^+, $CH_3OH_2^+$, and NH_4^+.[101] An interesting feature of the spectra obtained was the observation of exchange of the reagent gas for water or methanol of the dicarboxylic compound. Thus dodecanedioic acid on reaction with $CH_3OH_2^+$ showed MH^+ and both $[MH^+-H_2O+CH_3OH]$ and $[MH^+-2H_2O+2CH_3OH]$, with the latter ion being the most intense in the spectrum. Water elimination from $[MH^+-H_2O+CH_3OH]$ also was observed as was H_2O elimination from MH^+. These type of solvolyses reactions are not observed with the non-hydrogen bonding reagent gases CH_4 or $i-C_4H_{10}$.

The CH_4 CI mass spectra of a number of higher esters were reported in the early work of Munson and Field.[102] In addition, the H_2 and CH_4 CI mass spectra of a series of formate esters have been reported,[103] while the reaction of formate and acetate esters with a variety of Brønsted acids, with which the protonation exothermicity was systematically varied, has also been studied.[104] In addition to alcohol loss from MH^+ (Reaction 35), two new reaction channels, elimination of an alkene (Reaction 36) and elimination of a carboxylic acid (Reaction 37) became possible and energetically feasible. If the protonation step is sufficiently exothermic, further fragmentation of the alkyl ion R^+ may occur.[104]

$$\underset{\text{OH}}{R'C \overset{\displaystyle +}{\cdots} OR} \rightarrow R'C \equiv O^+ + ROH \tag{35}$$

$$\rightarrow \underset{\text{OH}}{R'-C \overset{\displaystyle +}{\cdots} OH} + (R-H) \tag{36}$$

$$\rightarrow R^+ + R'CO_2H \tag{37}$$

Figure 11 shows the CH_4 CI mass spectra of isopentyl and n-pentyl propionates. Moderate intensity (10% of base peak) MH^+ ions are observed with the main fragment ions being $C_2H_5C\equiv O^+$ (Reaction 35), $C_2H_5CO_2H_2^+$ (Reaction 36) and $C_5H_{11}^+$ (Reaction 37). The $C_2H_5^+$ and $C_3H_5^+$ reagent ions react in part by addition and elimination of [R−H], Reactions 38 and 39; these reactions are observed for propyl and higher esters

$$C_2H_5^+ + R'-CO_2R \rightarrow \underset{\displaystyle +}{R'-\underset{\text{OC}_2\text{H}_5}{C}-OR} \rightarrow \underset{\text{OH}}{R'-C \overset{\displaystyle +}{\cdots} OC_2H_5} + R-H \tag{38}$$

$$C_3H_5^+ + R'-CO_2R \rightarrow \underset{\displaystyle +}{R'-\underset{\text{OC}_3\text{H}_5}{C}-OR} \rightarrow \underset{\text{OH}}{R'-C \overset{\displaystyle +}{\cdots} OC_3H_5} + R-H \tag{39}$$

The results in Figure 11 show that formation of the acyl ion is of rather minor importance, the major fragmentation of MH^+ occurring by Reactions 36 or 37. For n-pentyl propionate $C_2H_5CO_2H_2^+$ (Reaction 36) is the base peak while for i-pentyl propionate $C_5H_{11}^+$ (Reaction 37) is the base peak. The competition between these two reactions is illustrated further by the results in Table 9 for the CO_2/H_2 CI of isomeric pentyl acetates[104] (CO_2H^+ has an acidity similar to that of CH_5^+). For those acetates containing an unbranched alkyl group, $CH_3CO_2H_2^+$ (Reaction 36) is the base peak while for those acetates containing a branched alkyl group the $C_5H_{11}^+$ alkyl ion (Reaction 37) is the base peak. Presumably, in the branched isomers, hydrogen shifts in the alkyl groups permit formation of a stable tertiary $C_5H_{11}^+$ ion, while for the linear isomers only the less stable secondary ions can be formed.

Finally, it should be noted that Field and co-workers[105-107] have observed that the extent of fragmentation of MH^+ in the i-C_4H_{10} CI of a number of acetates is strongly dependent on the temperature. This is to be expected whenever the protonation step is essentially thermoneutral. In such cases the thermal energy is comparable to the excess internal energy arising from the protonation step. When the protonation reaction is strongly exothermic the contribution of the thermal energy to the total excess internal energy will be negligible and a much smaller temperature dependence will be observed.

The OH^- CI mass spectra of carboxylic acids show $[M-H]^-$, presumably the carboxylate anion, as the sole product ion.[12] By contrast the esters studied[12] showed a much more complex behavior in their reaction with OH^-. While methyl, ethyl, and n-propyl acetates showed abundant $[M-H]^-$ ions the acetate ion, CH_3COO^-, increased in abundance with increasing size of the alkyl group ($[M-H]^-$, 42% and CH_3COO^-, 58% for n-propyl acetate). For t-pentyl acetate the alkoxy anion $C_2H_5(CH_3)_2CO^-$ dominated the spectrum (94% of total ionization). For methyl octanoate $[M-H]^-$, $RCOO^-$, and

113

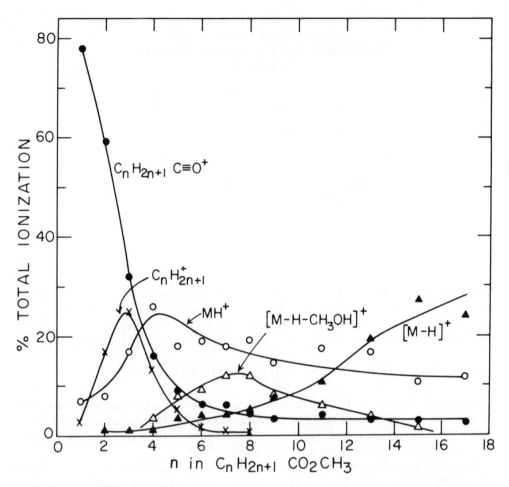

FIGURE 10. Dependence of ion abundances of chain length in H_2 CI of methyl n-alkanoates. (Reproduced, with permission, from *J. Chem. Soc. Perkin Trans. II*, 1718, 1975.)

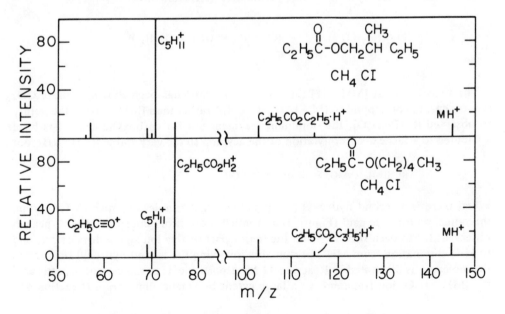

FIGURE 11. CH_4 CI mass spectra of isopentyl and n-pentyl propionates.

Table 9
RELATIVE FRAGMENT ION
INTENSITIES IN CO_2/H_2 CI OF PENTYL
ACETATES

Pentyl group	Intensity $C_5H_{11}^+$	Intensity $CH_3CO_2H_2^+$
$CH_3CH_2CH_2CH_2CH_2-$	45	100
$CH_3CH_2-CH-CH_2CH_3$	48	100
$CH_3CH_2CH_2CHCH_3$ $\quad\quad\quad CH_3$	44	100
$CH_3CHCH_2CH_2-$ $\quad CH_3$	100	57
$CH_3CH_2CHCH_2-$ $\quad\quad CH_3$	100	52
$CH_3-CH-CHCH_3$	100	7

CH_3O^- all were observed while for trimethylsilyl eicosanoate the carboxylate anion $RCOO^-$ (89%) and $[M-H]^-$ (10%) were the major products. In contrast to these results, Hunt et al.[108] have reported that only $[M-H]^-$ is produced in the reaction of OH^- with ethyl hexanoate, methyl crotonate, n-pentylacetate, and dimethyl suberate.

XIII. AMINO ACIDS AND DERIVATIVES

The H_2 CI[109] and CH_4 CI[109,110] mass spectra of a variety of amino acids have been determined. An additional study[111] has concentrated on the cluster ions $[2M+H]^+$, $[M+C_2H_5]^+$, $[M+C_3H_5]^+$ and their fragmentation modes. The i-C_4H_{10} CI mass spectra of a few amino acids also have been reported.[112]

The CH_4 CI mass spectra of three representative amino acids are presented in Figure 12. Unsubstituted amino acids, such as leucine, show relatively abundant MH^+ ions with the principle fragmentation reaction corresponding to elimination of the elements of formic acid (probably as $CO + H_2O$) from MH^+ (Reaction 40).

$$RCH(NH_2)CO_2H \cdot H^+ \rightarrow RCH=\overset{+}{N}H_2 + CO_2H_2 \ (CO + H_2O)$$

(40)

The weak signal at $[MH^+-18]$ (m/z 114 for leucine) has been shown,[109] using CD_4 reagent gas, to correspond to $[M+C_2H_4-CO_2H_2]^+$ rather than $[MH^+-H_2O]$; the failure to observed $[MH^+-H_2O]$, the dominant fragment ion for carboxylic acids, has been attributed to a facile decarbonylation of the acyl ion to the very stable $RCH=NH_2^+$ ion:

$$RCH(NH_2)C\equiv O^+ \rightarrow RCH=NH_2^+ + CO \qquad (41)$$

When there is a second hydroxyl group present in the molecule, such as for serine, threonine, or aspartic acid (Figure 12b) a much more pronounced $[MH^+-18]$ peak is observed. It has been shown[109] that the major part of this ion signal does correspond to H_2O elimination from MH^+, presumably involving the second hydroxyl function in the molecule as illustrated in Reaction 42 for aspartic acid. In the case of aspartic acid the $[MH^+-H_2O]$ ion fragments to a large extent by ketene elimination (Reaction 43).

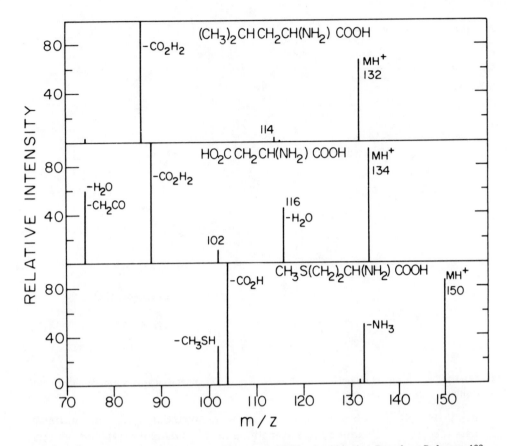

FIGURE 12. CH_4 CI mass spectra of leucine, aspartic acid, and methionine. Data from Reference 109.

(The ion signal at m/z 102 for aspartic acid corresponds to elimination of H_2O + CH_2CO from the $[M + C_2H_5]^+$ ion.) For serine and threonine sequential loss of two molecules of H_2O from MH^+ is observed.

$$HO \overset{+\,OH}{\overset{\Vert}{\underset{\quad}{C}}}-CH_2CH(NH_2)COOH \rightarrow \overset{+}{O}{\equiv}C-CH_2CH(NH_2)COOH + H_2O \qquad (42)$$

$$^+O{\equiv}C-CH_2CH(NH_2)COOH \rightarrow H_2\overset{+}{N}{=}CHCOOH + CH_2CO \qquad (43)$$

Loss of NH_3 from MH^+ is observed only when the amino acid contains a functional group which is capable of anchimerically assisting the elimination reaction as shown in Scheme 10 for methionine.

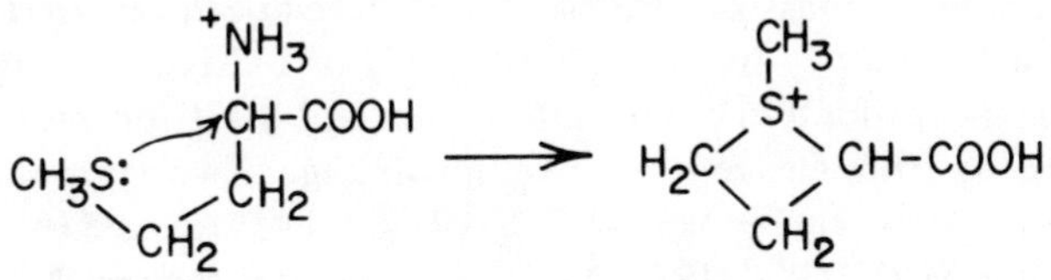

SCHEME 10

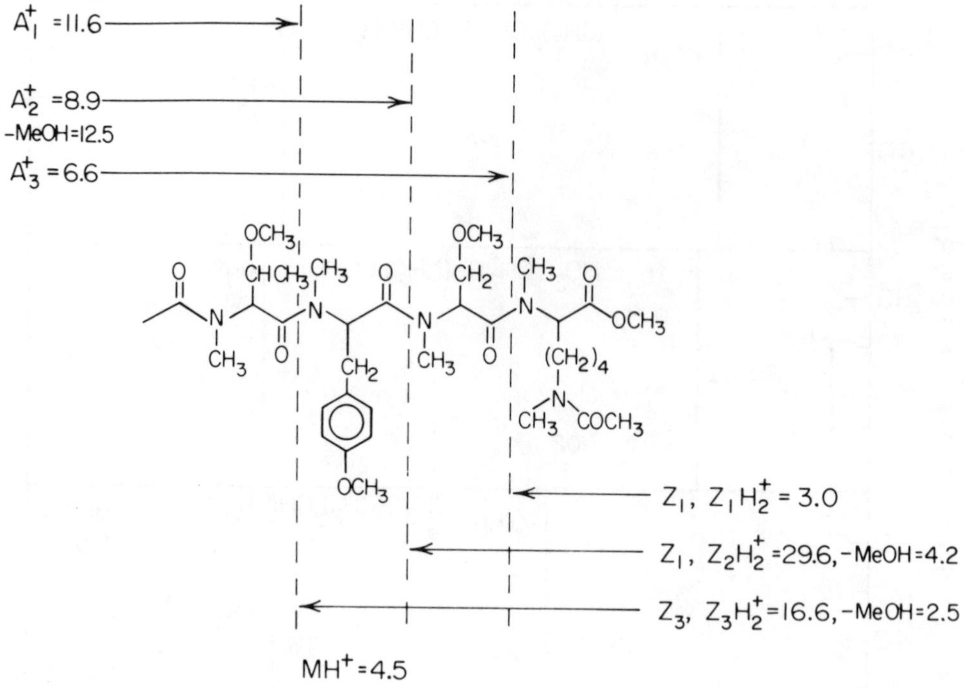

$A_1^+ = 11.6$

$A_2^+ = 8.9$
$-MeOH = 12.5$
$A_3^+ = 6.6$

$Z_1, Z_1H_2^+ = 3.0$

$Z_1, Z_2H_2^+ = 29.6, -MeOH = 4.2$

$Z_3, Z_3H_2^+ = 16.6, -MeOH = 2.5$

$MH^+ = 4.5$

FIGURE 13.   Schematic of fragmentation of an N-acetyl-N,O-permethylated tetrapeptide under i-C_4H_{10} CI conditions. Data from Reference 121.

Thus $[MH^+-NH_3]$ is observed in significant abundance only in the proton transfer CI mass spectra of such amino acids as methionine, phenylalanine, tyrosine, and tryptophan; indeed it becomes the base peak in the CH_4 CI mass spectrum of the latter.

The high MH^+ ion intensity in the CH_4 CI mass spectra of amino acids stands in contrast to the electron impact mass spectra where no molecular ions are observed.[113] The MH^+ ion intensity is much reduced following the more exothermic protonation by H_5^+ and there is a corresponding increase in the fragmentation, including more extensive sequential fragmentation processes. In the i-C_4H_{10} CI there is very little fragmentation of the MH^+ ion, although the extent of fragmentation increases markedly with increasing temperature.[112]

LeClerq and Desiderio[111] have reported the CH_4 CI mass spectra of a number of amino acid amides and methyl esters. Again abundant MH^+ ions are observed, with the major fragmentation routes being loss of NH_3 + CO for the amides and loss of CH_3OH + CO for the esters. The N^α-acetyl amino acid derivatives show abundant $[MH^+-CH_2CO]$ fragment ions as well as $[MH^+-CH_2CO-CO_2H_2]$ fragment ions.[111] Smit and Field[12] have reported the OH^- CI mass spectra of two amino acids, leucine and phenylalanine. The $[M-H]^-$ ion dominated the spectra with low-intensity $[M-3H]^-$ ions also being observed.

The proton transfer CI mass spectra of both underivatized and derivatized peptides have been studied extensively using primarily CH_4 or i-C_4H_{10} as reagent gases.[114-121] The CI mass spectra normally contain a low-intensity MH^+ ion plus more abundant fragment ions arising from cleavage at an amide linkage. Two main series of fragment ions arise, the N-terminus acyl ions $R'(NR''CHRCO)_n^+$ (A_n^+) formed by simple cleavage, and the C-terminus $H^+(NHR''CHRCO)_nOR''$ ($Z_nH_2^+$) ions formed by cleavage of the amide bond accompanied by hydrogen migration. (R' and R'' are either hydrogens or groups added during derivatization.) The type of spectrum obtained is illustrated schematically in Figure 13 for an N-acetyl-N,O-permethylated tetrapeptide under i-C_4H_{10} CI conditions.[121] Clearly, in favorable cases, considerable information concerning the

Table 10
CH$_4$ CI MASS SPECTRA OF C$_{27}$ STEROIDS

		Relative intensity			
Compound	MH$^+$	M$^{+\cdot}$	[M−H]$^+$	[MH$^+$−H$_2$O]	Other
5α-Cholestane	20	40	100	—	[MH$^+$−CH$_4$], 28
Cholest-5-ene	35	45	100	—	[MH$^+$−CH$_4$], 35
Cholest-3,5-diene	95	70	100	—	[MH$^+$−CH$_4$], 36
5α-Cholestan-3-one	100	15	35	—	
Cholest-4-ene-3-one	100	15	25	—	[MH$^+$−CH$_4$], 5
5α-Cholestan-3β-ol	5	10	25	100	—
Cholest-5-en-3β-ol	10	30	60	100	[M−H−H$_2$O]$^+$, 10
Cholest-5-en-3β,7α-diol	5	8	10	100	[M−H−H$_2$O]$^+$, 50
					[MH$^+$−2H$_2$O], 20

amino acid sequence can be derived from the complementary A$_n^+$ and Z$_n$H$_2^+$ series of ions. Collision-induced dissociation of the various A$_n^+$ and ZnH$_2^+$ ions also has been used to provide sequence information.[121]

It has been reported that the OH$^-$ CI mass spectra of underderivatized peptides show primarily [M−H]$^-$ ions.[122,123] Sequence information was obtained from the collision-induced dissociation of these [M−H]$^-$ ions. In contrast to this result is the reported CH$_3$O$^-$ CI mass spectrum of an N-acetyl-N,O-permethylated peptide in which an abundant [M−H]$^-$ ion was observed as well as fragment ions, CH$_3$CO(NMeCHRCONMe)$_n^-$, indicative of the amino acid sequence.[124]

XIV. STEROIDS

The CH$_4$ CI mass spectra of several steroids, primarily derivatives of 5α-androstane, have been reported.[125] In a more extensive study the CH$_4$, i-C$_4$H$_{10}$, and NH$_3$ CI mass spectra of 34 C$_{27}$-steroids were recorded.[126] The CH$_4$ CI mass spectra of a selection of these latter steroids are summarized in Table 10. The major ion species observed are the MH$^+$, [M−H]$^+$, and [MH$^+$−H$_2$O] ions, the latter being observed only when a hydroxyl group is present. In addition, for the alcohols [M−H−H$_2$O]$^+$ ions and for diols [MH$^+$−2H$_2$O] ions are observed. The M$^{+\cdot}$ ion intensities are much higher than those reported for the C$_{19}$-steroids[125] suggesting that there may be a significant electron impact contribution to the spectra. As the data show, the presence of a double bond or a carbonyl group enhances the MH$^+$ ion intensity while the presence of a hydroxyl substituent decreases the MH$^+$ ion intensity due to facile elimination of H$_2$O. Similar results were obtained in the earlier study.[125]

The i-C$_4$H$_{10}$ CI mass spectra are similar to the CH$_4$ CI mass spectra but show less abundant [M−H]$^+$ ion formation. The NH$_3$ CI mass spectra are considerably different, however, showing [M + NH$_4$]$^+$ and [MH]$^+$ formation as well as, for the hydroxy steroids, loss of H$_2$O from both [M + NH$_4$]$^+$ and MH$^+$. The [M + NH$_4$−H$_2$O]$^+$ ion is isobaric with M$^{+\cdot}$. The results show that a conjugated olefinic system, a conjugated ene-one system, or a conjugated dione system enhances MH$^+$ ion formation. This feature has been observed earlier in simple systems.[127] The spectra of several of the compounds were determined at both 100° and 195°C source temperature. The higher temperature resulted in more extensive fragmentation, particularly of the [M + NH$_4$]$^+$ ion.

The OH$^-$ CI mass spectra of the same series of C$_{27}$-steroids also have been determined.[128] A selection of the results obtained is presented in Table 11. The primary ionization reaction clearly leads to [M−H]$^-$ which may fragment by successive elimination of H$_2$, elimination of H$_2$O, or both. In several cases the [M−H]$^-$ forms an adduct ion [M + 43]$^-$ with the N$_2$O of the reagent gas mixture.

Table 11

OH⁻ CI MASS SPECTRA OF C_{27} STEROIDS

Compound	Relative intensity					
	$[M-H]^-$	$[M-3H]^-$	$[M-5H]^-$	$[M-H-H_2O]^-$	$[M-3H-H_2O]^-$	Other
5α-Cholestane	—	100	38	—	—	
Cholest-5-ene	13	75	50	—	—	$[M+43]^-$, 100
Cholest-3,5-diene	100	39	—	—	—	
5α-Cholestan-3-one	100	—	—	—	—	
Cholest-4-ene-3-one	100	8	—	—	—	
5α-Cholestan-3β-ol	100	35	—	15	6	$[M+43]^-$, 37
Cholest-5-en-3β-ol	100	—	—	14	17	$[M+43]^-$, 8
Cholest-5-en-3β,7α-diol	56	—	—	79	21	$[M-H-2H_2O]^-$, 100

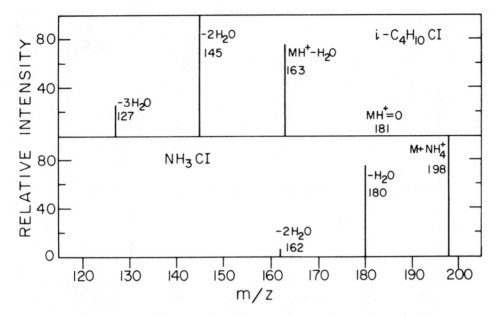

FIGURE 14. i-C$_4$H$_{10}$ and NH$_3$ CI mass spectra of D-glucopyranose. Data from Reference 41.

In the same study[128] the OH$^-$ CI mass spectra of a number of cholesteryl esters were recorded. The major product ions observed were [M−H]$^-$ and the carboxylate anion RCOO$^-$ derived from the acid function. For cholesteryl benzoate the carboxylate ion was the only product observed.

XV. CARBOHYDRATES AND DERIVATIVES

The electron impact mass spectra of carbohydrates and derivatives frequently result from complex fragmentation paths and show very low molecular ion intensities.[129,130] The first indication of the potential of chemical ionization in the analysis of carbohydrates came with the publication of the CH$_4$ CI mass spectrum of 2-deoxy-D-ribose;[131] although the MH$^+$ ion intensity was extremely small a simple fragmentation pattern resulting from sequential loss of up to three molecules of H$_2$O from MH$^+$ was observed. Subsequently a variety of carbohydrates have been examined by CI using primarily proton transfer reagent ions.[40,41,132-135]

Chemical ionization using NH$_3$[40,41] or NH$_3$/i-C$_4$H$_{10}$ mixtures[133] is most suitable for molecular weight determination since intense [M + NH$_4$]$^+$ ions are formed in most cases, including oligosaccharide peracetates.[133] The NH$_3$/i-C$_4$H$_{10}$ mixture gives NH$_4^+$ as the reactant ion. The use of CH$_4$ or pure i-C$_4$H$_{10}$ as proton transfer reagent systems normally results in very low MH$^+$ ion abundances but does give fragmentation sequences which provide structural information. For example, Figure 14 shows the NH$_3$ and i-C$_4$H$_{10}$ CI mass spectra of D-glucopyranose while Figure 15 shows the NH$_3$ and i-C$_4$H$_{10}$ CI mass spectra of methyl β-glucopyranoside tetraacetate.[41] The NH$_3$ CI mass spectrum of D-glucopyranose shows [M + NH$_4$]$^+$ as the base peak with significant loss of H$_2$O from this adduct ion. In contrast, the i-C$_4$H$_{10}$ CI mass spectrum shows no MH$^+$ ion, only fragment ions corresponding to sequential loss of one, two, and three molecules of water from MH$^+$. Using isotopic labeling, the sequence of fragmentation reactions in the CH$_4$ CI mass spectrum of D-glucopyranose pentacetate was found[40] to be as shown in Scheme 11. The specificity of initial loss of the CI substituent also holds in the NH$_3$ and i-C$_4$H$_{10}$ CI mass spectra of methyl β-D-glucopyranoside tetraacetate. A weak [MH$^+$−MeOH] fragment is observed in the NH$_3$CI spectrum while the i-C$_4$H$_{10}$ CI mass spectrum can best be rationalized in terms of initial loss of CH$_3$OH followed by further elimination of acetic acid (HOAc).

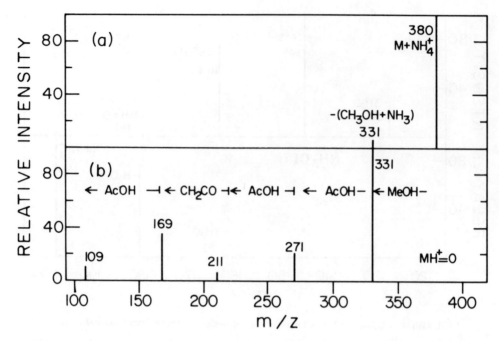

FIGURE 15.   NH₃ CI (a) and i-C₄H₁₀ (b) mass spectra of methyl β-glucopyranose tetraacetate. Data from Reference 41.

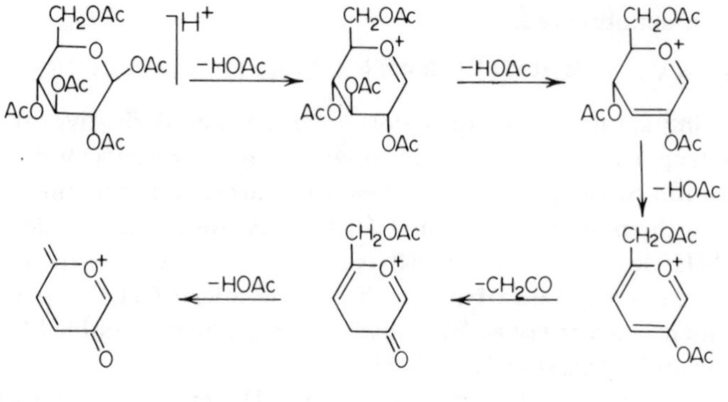

SCHEME 11

Horton et al.[41] have examined the NH₃ and i-C₄H₁₀ CI of a variety of carbohydrate derivatives. They found that in the NH₃ CI of dithioacetal derivatives the [M + NH₄]⁺ ion fragmented by sequential loss of two molecules of the appropriate alkanethiol. In the i-C₄H₁₀ CI initial fragmentation of MH⁺ (not observed) was by loss of a neutral alkanethiol followed by loss of H₂O for the free sugars or loss of HOAc for the acetate derivatives; further loss of an alkanethiol and H₂O/HOAc also occurred. The NH₃ CI of diisopropylidene derivatives showed both [M + NH₄]⁺ and MH⁺ ions and sequential loss of two (CH₃)₂CO molecules from both. The i-C₄H₁₀ CI showed no MH⁺ ion signals but showed fragment ions corresponding to sequential loss of (CH₃)₂CO and H₂O from MH⁺.

Dougherty and co-workers[133] have reported the CI mass spectra of a number of oligosaccharide peracetates using an i-C₄H₁₀/NH₃ mixture as reagent gas. Intense ions corresponding to [M + NH₄]⁺ were observed for all but the pentasaccharides. Ions which corresponded to thermolysis fragments generally dominated the fragmentation pattern. In later work Dougherty and co-workers[135] reported the i-C₄H₁₀ and i-C₄H₁₀/

NH$_3$ CI mass spectra of six permethylated glucosylalditols and two permethylated biosylalditols. Intense peaks corresponding, respectively, to MH$^+$ and [M + NH$_4$]$^+$ were observed. The sequence of monosaccharide units could be determined from the fragment ions in the i-C$_4$H$_{10}$ CI mass spectra.

Ganguly et al.[136] have reported the Cl$^-$ CI mass spectra of two oligosaccharides. They observed [M + Cl]$^-$ as the base peak for both compounds. The major fragment ions corresponded to the loss of one or more discrete sugar units from either end of the oligosaccharide chain. All cleavages involved glycosidic linkages and only Cl$^-$ adduct ions were sufficiently intense to be recorded. They noted that the presence of free hydroxyl groups was necessary for Cl$^-$ attachment to occur and that permethylated sugars gave poor negative ion CI mass spectra.

The proton transfer CI mass spectra of glucoronides (Structure A) and derivatives show very low intensities for ions (MH$^+$, [M + NH$_4$]$^+$) from which the molecular weight can be derived; the fragment ions which are seen derive primarily from the carbohydrate portion of the molecule.[137-139] Fenselau and co-workers[139,140] have shown that the addition of pyridine to i-C$_4$H$_{10}$ or NH$_3$ gives abundant protonated pyridine ions which form stable adduct ions with pertrimethylsilyl glucoronides derivatives thus permitting molecular weights to be established.

Recently Bruins[141] has reported the OH$^-$ mass spectra of a number of underivatized glucoronides using desorption chemical ionization techniques. For most of the examples studied the [M−H]$^-$ ion was the base peak with a significant peak corresponding to RO$^-$. In addition, for a number of steroidal glucoronides an ion was observed corresponding to elimination of H$_2$O from [M−H]$^-$. The one exception to this pattern occurred when R was p-nitrophenyl. In this case M$^-$ and [M−OH]$^-$ as well as RO$^-$ were observed. It appears that quasi-thermal energy electrons were present which reacted with the nitro compound by electron attachment at a much faster rate than OH$^-$ reacted.

STRUCTURE A   STRUCTURE B

McCloskey and co-workers[142] have shown that the CH$_4$ CI mass spectra of nucleosides (Structure B), consisting of a sugar moiety and a nitrogen base B, show abundant MH$^+$ ions with the fragment ion BH$_2^+$ usually being the base peak in the spectrum. Weaker signals arising from the sugar moiety are observed. The high MH$^+$ ion abundance is due to the strongly basic nitrogen centers present in B. In a more recent study[143] MH$^+$ and BH$_2^+$ ions were observed in the NH$_3$ CI mass spectra of several isomeric 7- and 9-β-D-ribofuranosyl purines. Wilson and McCloskey[144] have studied the proton transfer CI mass spectra of a number of nucleosides using CH$_4$, i-C$_4$H$_{10}$, NH$_3$, CH$_3$NH$_2$, (CH$_3$)$_2$NH, and (CH$_3$)$_3$N as reagent gases. When proton transfer is not exothermic, adducts of the reagent ion and the nucleosides were observed. From the observance or nonobservance of proton transfer from the different reagent ions approximate proton affinities of the nucleosides were derived. Recently, the NH$_3$CI mass spectra of a number of purine and pyrimidine nucleosides have been obtained[145] using desorption chemical ionization techniques; MH$^+$, [M + NH$_4$]$^+$, BH$_2^+$, [BH + NH$_4$]$^+$, [S−H]$^+$, and [S−H + NH$_4$]$^+$ ions were observed. Of particular note are the good CI mass spectra of cyclic adenosine monophosphate which have been obtained[145,146] by desorption chemical ionization methods.

REFERENCES

1. Field, F. H., Munson, M. S. B., and Becker, D. A., Chemical ionization mass spectrometry, *Adv. Chem. Ser.*, 58, 167, 1966.
2. Clow, R. P. and Futrell, J. H., Ion cyclotron resonance study of the mechanism of chemical ionization. Mass spectroscopy of selected hydrocarbons using methane reagent gas, *J. Am. Chem. Soc.*, 94, 3748, 1972.
3. Houriet, R., Parisod, G., and Gaümann, T., The mechanism of chemical ionization of n-paraffins, *J. Am. Chem. Soc.*, 99, 3599, 1977.
4. Hiroaka, K. and Kebarle, P., Stability and energetics of penta-coordinated carbonium ions. The isomeric $C_2H_7^+$ ions and some higher analogues: $C_3H_9^+$ and $C_4H_{11}^+$, *J. Am. Chem. Soc.*, 98, 6119, 1976.
5. Hunt, D. F. and McEwen, C. N., Chemical ionization mass spectrometry. VII. Deuterium-labeled decanes, *Org. Mass Spectrom.*, 7, 441, 1973.
6. Bowen, R. D. and Williams, D. H., The concept of a hierarchy of unimolecular reactions in a homologous series. Prediction of the unimolecular chemistry of some saturated carbenium ions, *J. Chem. Soc. Perkin Trans. II*, 1479, 1976.
7. Hunt, D. F. and Harvey, T. M., Nitric oxide chemical ionization mass spectra of alkanes, *Anal. Chem.*, 47, 1965, 1975.
8. Hunt, D. F., McEwen, C. N., and Harvey, T. M., Positive and negative chemical ionization mass spectrometry using a Townsend discharge ion source, *Anal. Chem.*, 47, 1730, 1975.
9. Field, F. H., Chemical ionization mass spectrometry. VIII. Alkenes and alkynes, *J. Am. Chem. Soc.*, 90, 5649, 1968.
10. Budzikiewicz, H. and Busker, E., Studies in chemical ionization mass spectrometry. III. CI-spectra of olefins, *Tetrahedron*, 36, 255, 1980.
11. Hunt, D. F. and Harvey, T. M., Nitric oxide chemical ionization mass spectra of olefins, *Anal. Chem.*, 47, 2136, 1975.
12. Smit, A. L. C. and Field, F. H., Gaseous anion chemistry. Formation and reaction of OH^-. Reactions of anions with N_2O, OH^- negative chemical ionization, *J. Am. Chem. Soc.*, 99, 6471, 1977.
13. Bohme, D. K. and Young, L. B., Gas phase reactions of oxide radical ion and hydroxide ion with simple olefins and of carbanions with oxygen, *J. Am. Chem. Soc.*, 92, 3301, 1970.
14. Blum, W. and Richter, W. J., Analyses of alkene mixtures by combined capillary gas chromatography-chemical ionization mass spectrometry, *Tetrahedron Lett.*, 835, 1973.
15. Ferrer-Correia, A. J., Jennings, K. R., and Sen Sharma, D. K., The use of ion-molecule reactions in the mass spectrometric location of double bonds, *Org. Mass Spectrom.*, 11, 867, 1976.
16. Ferrer-Correia, A. J., Jennings, K. R., and Sen Sharma, D. K., The use of ion-molecule reactions in the mass spectrometric location of double bonds, *Adv. Mass Spectrom.*, 7, 287, 1978.
17. Greathead, R. J. and Jennings, K. R., The location of double bonds in mono- and di-unsaturated compounds, *Org. Mass Spectrom.*, 15, 431, 1980.
18. Chai, R. and Harrison, A. G., Location of double bonds by chemical ionization mass spectrometry, *Anal. Chem.*, 53, 34, 1981.
19. Field, F. H. and Munson, M. S. B., Chemical ionization mass spectrometry. V. Cycloparaffins, *J. Am. Chem. Soc.*, 89, 4272, 1967.
20. Busker, E. and Budzikiewicz, H., Studies in chemical ionization mass spectrometry. 2. $i-C_4H_{10}$ and NO spectra of alkynes, *Org. Mass Spectrom.*, 14, 222, 1979.
21. Munson, M. S. B. and Field, F. H., Chemical ionization mass spectrometry. IV. Aromatic hydrocarbons, *J. Am. Chem. Soc.*, 89, 1047, 1967.
22. Field, F. H., Chemical ionization mass spectrometry. VI. C_7H_8 isomers, toluene, cycloheptatriene, and norbornadiene, *J. Am. Chem. Soc.*, 89, 5328, 1967.
23. Leung, H.-W. and Harrison, A. G., Hydrogen migrations in mass spectrometry. IV. Formation of $C_6H_7^+$ in the chemical ionization mass spectra of alkylbenzenes, *Org. Mass Spectrom.*, 12, 582, 1977.
24. Harrison, A. G., Lin, P.-H., and Leung, H.-W., The chemical ionization of alkylbenzenes, *Adv. Mass Spectrom.*, 7, 1394, 1978.
25. Wesdemiotis, C., Schwarz, H., Van de Sande, C. C., and Van Gaever, F., Transalkylation and proton-catalyzed C−C cleavage of gaseous n-butyl and n-pentyl benzenes, *Z. Naturforsch.*, 34b, 495, 1979.
26. Herman, J. A. and Harrison, A. G., Effect of protonation exothermicity on the chemical ionization mass spectra of some alkylbenzenes, *Org. Mass Spectrom.*, 16, 425, 1981.
27. Hatch, F. and Munson, B., Reactant ion monitoring for selective detection in gas chromatography/chemical ionization mass spectrometry, *Anal. Chem.*, 49, 731, 1977.
28. Einolf, N. and Munson, B., High pressure charge exchange mass spectrometry, *Int. J. Mass Spectrom. Ion Phys.*, 9, 141, 1972.
29. Hunt, D. F., Reagent gases for chemical ionization mass spectrometry, *Adv. Mass Spectrom.*, 6, 517, 1974.

30. Bruins, A. P., Ferrer-Correia, A. J., Harrison, A. G., Jennings, K. R., and Mitchum, R. K., Negative ion chemical ionization mass spectra of some aromatic compounds using O⁻ as the reagent ion, *Adv. Mass Spectrom.*, 7, 355, 1978.

31. Jennings, K. R., Investigations of selective reagent ions in chemical ionization mass spectrometry, in *High Performance Mass Spectrometry*, Gross, M. L., Ed., American Chemical Society, Washington, D.C., 1978.

32. Jennings, K. R., Negative chemical ionization mass spectrometry, in *Mass Spectrometry*, Vol. 4, Spec. Period. Rep., Chemical Society, London, 1977.

33. Field, F. H., Chemical ionization mass spectrometry. XII. Alcohols, *J. Am. Chem. Soc.*, 92, 2672, 1970.

34. Ichikawa, H. and Harrison, A. G., Hydrogen migrations in mass spectrometry. VI. The chemical ionization mass spectra of benzoic acids and benzyl alcohols, *Org. Mass Spectrom.*, 13, 389, 1978.

35. Herman, J. A. and Harrison, A. G., Effect of reaction exothermicity on the proton transfer chemical ionization mass spectra of isomeric C_5 and C_6 alkanols, *Can. J. Chem.*, 59, 2125, 1981.

36. Munson, M. S. B. and Field, F. H., Chemical ionization mass spectrometry. I. General introduction, *J. Am. Chem. Soc.*, 88, 2621, 1966.

37. Dzidic, I. and McCloskey, J. A., Influence of remote functional groups on the chemical ionization mass spectra of long-chain compounds, *J. Am. Chem. Soc.*, 93, 4955, 1971.

38. Winkler, J. and McLafferty, F. W., Stereochemical effects in the chemical ionization mass spectra of cyclic diols, *Tetrahedron*, 30, 2971, 1974.

39. Munson, B., Chemical ionization mass spectrometry: analytical applications of ion-molecule reactions, in *Interactions Between Ions and Molecules*, Ausloos, P., Ed., Plenum Press, New York, 1975.

40. Hogg, A. M. and Nagabhusan, T. L., Chemical ionization mass spectra of sugars, *Tetrahedron Lett.*, 4827, 1972.

41. Horton, D. B., Wander, J. D., and Foltz, R. L., Analysis of sugar derivatives by chemical ionization mass spectrometry, *Carbohyd. Res.*, 36, 75, 1974.

42. Hunt, D. F. and Ryan, J. F., Chemical ionization mass spectrometry studies: nitric oxide as a reagent gas, *J. Chem. Soc. Chem. Commun.*, 620, 1972.

43. Houriet, R., Stahl, D., and Winkler, F. J., Negative chemical ionization of alcohols, *Environ. Health Perspect.*, 36, 63, 1980.

44. Audier, H. E., Millet, A., Perret, C., and Varenne, P., Mecanismes de fragmentation en spectrometrie de masse par ionization chimique. IV. Fragmentation des iones oxiranes protonés, *Org. Mass Spectrom.*, 14, 129, 1979.

45. Bowen, R. D. and Harrison, A. G., Chemical ionization mass spectra of selected C_4H_8O compounds, *J. Chem. Soc. Perkin Trans. II*, 1544, 1981.

46. Benoit, F. M. and Harrison, A. G., Hydrogen migrations in mass spectrometry. I. Loss of olefin from phenyl n-propyl ether following electron impact and chemical ionization, *Org. Mass Spectrom.*, 11, 599, 1976.

47. Bowen, R. D. and Harrison, A. G., Chemical ionization mass spectra of selected C_3H_6O compounds, *Org. Mass Spectrom.*, 16, 159, 1981.

48. Fales, H. M., Fenselau, C., and Duncan, J. H., The loss of water from isotopically labeled heptanals in chemical ionization, *Org. Mass Spectrom.*, 11, 669, 1976.

49. Jardine, I. and Fenselau, C., The high pressure nitric oxide mass spectra of aldehydes, *Org. Mass Spectrom.*, 10, 748, 1975.

50. Harrison, A. G. and Jennings, K. R., Reactions of O⁻ with carbonyl compounds, *J. Chem. Soc. Faraday Trans. II*, 72, 1601, 1976.

51. Dawson, J. H. J., Noest, A. J., and Nibbering, N. M. M., 1,1- and 1,3-Elimination of water from the reaction complex of O⁻ with 1,1,1-trideuteroacetone, *Int. J. Mass Spectrom. Ion Phys.*, 30, 189, 1979.

52. Colosimo, M., Bucci, R., and Brancaleoni, E., The stability of alkylchloronium ions during H_3^+ CIMS of simple chloroalkanes, *Int. J. Mass Spectrom. Ion Phys.*, 39, 145, 1981.

53. Harrison, A. G., Lin, P.-H., and Tsang, C. W., Proton transfer reactions by trapped ion mass spectrometry, *Int. J. Mass Spectrom. Ion Phys.*, 19, 23, 1976.

54. Jardine, I. and Fenselau, C., Proton localization in chemical ionization fragmentation, *J. Am. Chem. Soc.*, 98, 5086, 1976.

55. Harrison, A. G. and Lin, P.-H., The chemical ionization mass spectra of fluorotoluenes, *Can. J. Chem.*, 53, 1314, 1975.

56. Leung, H.-W. and Harrison, A. G., Structural and energetics effects in the chemical ionization of halogen-substituted benzenes and toluenes, *Can. J. Chem.*, 54, 3439, 1976.

57. Leung, H.-W., Ichikawa, H., Li, Y.-H., and Harrison, A. G., Concerning the mechanism of dehalogenation of halobenzene derivatives by gaseous Brønsted acids, *J. Am. Chem. Soc.*, 100, 2479, 1978.

58. Leung, H.-W. and Harrison, A. G., The role of energetics in the hydrogen chemical ionization of halobenzene derivatives. Estimates of the heats of formation of substituted phenyl cations, *J. Am. Chem. Soc.*, 101, 3168, 1979.

59. Leung, H.-W. and Harrison, A. G., Specific substituent effects in the dehalogenation of halobenzene derivatives by the gaseous Brønsted acid CH$_5^+$, *J. Am. Chem. Soc.*, 102, 1623, 1980.

60. Liauw, W. G., Lin, M. S., and Harrison, A. G., Effect of protonation exothermicity on the reaction of gaseous Brønsted acids with halobenzene derivatives, *Org. Mass Spectrom.*, 16, 383, 1981.

61. Liauw, W. G. and Harrison, A. G., Site of protonation in the reaction of gaseous Brønsted acids with halobenzene derivatives, *Org. Mass Spectrom.*, 16, 388, 1981.

62. Cacace, F. and Speranza, M., Aromatic substitution in the gas phase. Ambient behavior of halo- and dihalo-benzenes towards D$_2$T$^+$. Tritio-deprotonation and tritio-dehalogenation, *J. Am. Chem. Soc.*, 98, 7299, 1976.

63. Speranza, M. and Cacace, F., Aromatic substitution in the gas phase. On the mechanism of the dehalogenation reactions of halobenzenes and dihalobenzenes promoted by gaseous Brønsted acids, *J. Am. Chem. Soc.*, 99, 3051, 1977.

64. Speranza, M., Sefcik, M. D., Henis, J. M. S., and Gaspar, P. P., Phenylium (C$_6$H$_5^+$) ion-molecule reactions studied by ion cyclotron resonance spectroscopy, *J. Am. Chem. Soc.*, 99, 5583, 1977.

65. Rosenstock, H. M., Draxl, K., Steiner, B. W., and Herron, J. T., Energetics of gaseous ions, *J. Phys. Chem. Ref. Data*, 6(1), 1977.

66. Rosenstock, H. M., Larkins, J. T., and Walker, J. A., Interpretation of photoionization thresholds. Quasiequilibrium theory and the fragmentation of benzene, *Int. J. Mass Spectrom. Ion Phys.*, 11, 309, 1973.

67. Harrison, A. G. and Kallury, R. K. M. R., The chemical ionization mass spectra of mononitroarenes, *Org. Mass Spectrom.*, 15, 284, 1980.

68. Harrison, A. G., Onuska, F. I., and Tsang, C. W., Chemical ionization mass spectra of selected polychlorinated biphenyl isomers, *Anal. Chem.*, 53, 1183, 1981.

69. Biros, F. J., Dougherty, R. C., and Dalton, J., Positive chemical ionization mass spectra of polycyclic aromatic pesticides, *Org. Mass Spectrom.*, 6, 1161, 1972.

70. McKinney, J. D., Oswald, E. O., Palaszek, S. M., and Corbett, B. J., Characterization of chlorinated hydrocarbon pesticide metabolites of the DDT and polycyclodiene types by electron impact and chemical ionization mass spectrometry, in *Mass Spectrometry and NMR Spectroscopy in Pesticide Chemistry*, Biros, F. and Haque, R., Eds., Plenum Press, New York, 1974.

71. Oswald, E. O., Albro, P. W., and McKinney, J. D., Use of gas-liquid chromatography coupled with chemical ionization and electron impact mass spectrometry for the investigation of potentially hazardous environmental agents and their metabolites, *J. Chromatogr.*, 98, 363, 1974.

72. Dougherty, R. C., Roberts, J. D., and Biros, F. J., Positive and negative chemical ionization mass spectra of some aromatic chlorinated pesticides, *Anal. Chem.*, 47, 54, 1975.

73. Crow, F. W., Bjorseth, A., Knapp, K. T., and Bennett, R., Determination of polyhalogenated hydrocarbons by glass capillary gas chromatography-negative ion chemical ionization mass spectrometry, *Anal. Chem.*, 53, 619, 1981.

74. Cairns, T. C. and Siegmund, E. G., Determination of polychlorinated biphenyls by chemical ionization mass spectrometry, *Anal. Chem.*, 53, 1599, 1981.

75. Dzidic, I., Carroll, D. I., Stillwell, R. N., and Horning, E. C., Atmospheric pressure ionization (API) mass spectrometry: formation of phenoxide ions from chlorinated aromatic compounds, *Anal. Chem.*, 47, 1308, 1975.

76. Busch, K. L., Hass, J. R., and Bursey, M. M., The gas enhanced negative ion mass spectra of polychloroanisoles, *Org. Mass Spectrom.*, 13, 604, 1978.

77. Busch, K. L., Norström, Å., Bursey, M. M., Hass, J. R., and Nelsson, C. A., Methane-oxygen enhanced negative ion mass spectra of polychlorinated diphenyl ethers, *Biomed. Mass Spectrom.*, 6, 157, 1979.

78. Busch, K. L., Norström, Å., Nelsson, C. A., Bursey, M. M., and Hass, J. R., Negative ion mass spectra of some polychlorinated 2-phenoxy-phenols, *Environ. Health Perspect.*, 36, 125, 1980.

79. Hunt, D. F., Harvey, T. M., and Russell, J. W., Oxygen as a reagent gas for the analysis of 2,3,7,8-tetrachlorodibenzo-p-dioxin by negative chemical ionization mass spectrometry, *J. Chem. Soc. Chem. Commun.*, 151, 1975.

80. Hass, J. R., Friesen, M. D., Harvin, D. J., and Parker, C. E., Determination of chlorinated dibenzo-p-dioxins in biological samples by negative chemical ionization mass spectrometry, *Anal. Chem.*, 50, 1474, 1978.

81. Mitchum, R. K., Althaus, J. R., Korfmacher, W. A., and Moler, G. F., Determination of polychlorinated dibenzo-p-dioxins in biological samples by negative chemical ionization mass spectrometry, *Adv. Mass Spectrom.*, 8, 1415, 1980.

82. Dougherty, R. C., Dalton, J., and Biros, F. J., Negative chemical ionization mass spectra of polycyclic chlorinated compounds, *Org. Mass Spectrom.*, 6, 1171, 1972.

83. Tannenbaum, H. P., Roberts, J. D., and Dougherty, R. C., Negative chemical ionization mass spectrometry-chloride attachment spectra, *Anal. Chem.*, 47, 49, 1975.

84. Audier, H. E., Millet, A., Perret, C., Tabet, J. C., and Varenne, P., Mecanismes de fragmentation en spectrometrie de mass par ionisation chimique. III. Ions formés par cyclisation, *Org. Mass Spectrom.*, 13, 315, 1978.

85. Davis, D. V. and Cooks, R. G., Site of protonation and bifunctional group interactions in α,ω-hydroxyalkylamines, *Org. Mass Spectrom.*, 16, 176, 1981.

86. Longevialle, P., Milne, G. W. A., and Fales, H. M., Chemical ionization mass spectrometry of complex molecules. XI. Stereochemical and conformational effects in the isobutane chemical ionization mass spectra of some steroidal amino alcohols, *J. Am. Chem. Soc.*, 95, 6666, 1973.

87. Longevialle, P., Girard, J.-P., Rossi, J.-C., and Tichy, M., Influence of the interfunctional distance on the dehydration of amino alcohols in isobutane chemical ionization mass spectrometry, *Org. Mass Spectrom.*, 14, 414, 1979.

88. Longevialle, P., Girard, J.-P., Rossi, J.-C., and Tichy, M., Isobutane chemical ionization mass spectrometry of β-aminoalcohols. Evaluation of population of conformers at equilibrium in the gas phase, *Org. Mass Spectrom.*, 15, 268, 1980.

89. Whitney, T. A., Klemann, L. P., and Field, F. H., Investigation of polytertiary alkylamines using chemical ionization mass spectrometry, *Anal. Chem.*, 43, 1048, 1971.

90. Maquestiau, A., Van Haverbeke, Y., Flammang, R., Mispreuve, H., and Elguero, J., Ionisation chimique de composés aromatiques nitres, *Org. Mass Spectrom.*, 14, 117, 1979.

91. Brophy, J. J., Diakiw, V., Goldsack, R. J., Nelson, D., and Shannon, J. S., Anomalous ions in the chemical ionization mass spectra of aromatic nitro and nitroso compounds, *Org. Mass Spectrom.*, 14, 201, 1979.

92. Zitrin, Y. and Yinon, J., Chemical ionization mass spectra of 2,4,6-trinitroaromatic compounds, *Org. Mass Spectrom.*, 11, 388, 1976.

93. Yinon, J. and Laschever, M., Reduction of trinitroaromatic compounds in water by chemical ionization mass spectrometry, *Org. Mass Spectrom.*, 16, 264, 1981.

94. Chizov, O. S., Kadentsev, V. I., Plambach, G. G., Bursten, K. I., Shevelev, S. A., and Fensilberg, A. A., Chemical ionization of aliphatic nitro compounds, *Org. Mass Spectrom.*, 13, 611, 1978.

95. Weinkam, R. J. and Gal, J., Effects of bifunctional interactions on the chemical ionization mass spectrometry of carboxylic acids and methyl esters, *Org. Mass Spectrom.*, 11, 188, 1976.

96. Harrison, A. G. and Kallury, R. K. M. R., Stereochemical applications of mass spectrometry. II. Chemical ionization mass spectra of isomeric dicarboxylic acids and derivatives, *Org. Mass Spectrom.*, 15, 277, 1980.

97. Benoit, F. M. and Harrison, A. G., Predictive value of proton affinity-ionization energy correlations involving oxygen-containing molecules, *J. Am. Chem. Soc.*, 99, 3980, 1977.

98. Middlemiss, N. E. and Harrison, A. G., Structure and fragmentation of gaseous protonated acids, *Can. J. Chem.*, 57, 2827, 1979.

99. Harrison, A. G. and Ichikawa, H., Site of protonation in the chemical ionization mass spectra of olefinic methyl esters, *Org. Mass Spectrom.*, 15, 244, 1980.

100. Tsang, C. W. and Harrison, A. G., Effect of chain length on the chemical ionization mass spectra of methyl n-alkanoates, *J. Chem. Soc. Perkin Trans. II*, 1718, 1975.

101. Weinkam, R. J. and Gal, J., Hydrolysis, methanolysis, and ammonolysis of dicarboxylic acids and methyl esters under conditions of chemical ionization, *Org. Mass Spectrom.*, 11, 197, 1976.

102. Munson, M. S. B. and Field, F. H., Chemical ionization mass spectrometry, II. Esters, *J. Am. Chem. Soc.*, 88, 4337, 1966.

103. Harrison, A. G. and Tsang, C. W., The hydrogen and methane chemical ionization mass spectra of some formate esters, *Can. J. Chem.*, 54, 2029, 1976.

104. Herman, J. A. and Harrison, A. G., Energetics and structural effects in the fragmentation of protonated esters in the gas phase, *Can. J. Chem.*, 59, 2133, 1981.

105. Field, F. H., Chemical ionization mass spectrometry. IX. Temperature and pressure studies with benzylacetate and t-amylacetate, *J. Am. Chem. Soc.*, 91, 2837, 1969.

106. Laurie, W. A. and Field, F. H., Chemical ionization mass spectrometry. XVI. Temperature effects in tertiary alkyl acetates, *J. Am. Chem. Soc.*, 94, 2913, 1972.

107. Laurie, W. A. and Field, F. H., Chemical ionization mass spectrometry. XVII. Effect of acid identity on decomposition rates of protonated tertiary alkyl esters, *J. Am. Chem. Soc.*, 94, 3359, 1972.

108. Hunt, D. F., Shabanowitz, J., and Giordani, A. B., Collision activated decomposition of negative ions in mixture analysis with a triple quadrupole mass spectrometer, *Anal. Chem.*, 52, 386, 1980.

109. Tsang, C. W. and Harrison, A. G., The chemical ionization of amino acids, *J. Am. Chem. Soc.*, 98, 1301, 1976.

110. Milne, G. W. A., Axenrod, T., and Fales, H. M., Chemical ionization mass spectrometry of complex molecules. IV. Amino acids, *J. Am. Chem. Soc.*, 92, 5170, 1970.

111. LeClerq, P. A. and Desiderio, D. M., Chemical ionization mass spectra of amino acids and derivatives. Occurrence and fragmentation of ion-molecule reaction products, *Org. Mass Spectrom.*, 7, 515, 1973.

112. Meot-Ner, M. and Field, F. H., Chemical ionization mass spectrometry. XX. Energy effects and virtual ion temperature in the decomposition of amino acids and amino acid derivatives, *J. Am. Chem. Soc.*, 95, 7207, 1973.

113. Junk, G. and Svec, H., The mass spectra of the α-amino acids, *J. Am. Chem. Soc.*, 85, 839, 1963.

114. Gray, W. R., Wojcik, L. H., and Futrell, J. H., Application of mass spectrometry to protein chemistry. II. Chemical ionization studies on acetylated permethylated peptides, *Biochem. Biophys. Res. Commun.*, 41, 1111, 1970.

115. Kiryuskin, A. A., Fales, H. M., Axenrod, T., Gilbert, E. J., and Milne, G. W. A., Chemical ionization mass spectrometry of complex molecules. VI. Peptides, *Org. Mass Spectrom.*, 5, 19, 1971.

116. Bowen, D. V. and Field, F. H., Isobutane chemical ionization mass spectrometry of dipeptides, *Int. J. Pept. Protein Res.*, 5, 435, 1973.

117. Baldwin, M. A. and McLafferty, F. W., Direct chemical ionization of relatively involatile samples. Application to underivatized oligopeptides, *Org. Mass Spectrom.*, 7, 1353, 1973.

118. Beuhler, R. J., Flanigan, E., Greene, L. J., and Friedman, L., Proton transfer mass spectrometry of peptides. A rapid heating technique for underivatized peptides containing arginine, *J. Am. Chem. Soc.*, 96, 3990, 1974.

119. Mudgett, M., Bowen, D. V., Kendt, T. J., and Field, F. H., C-Methylation: an artifact in peptides derivatized for sequencing by mass spectrometry, *Biomed. Mass Spectrom.*, 2, 254, 1975.

120. Mudgett, M., Bowen, D. V., Field, F. H., and Kendt, T. J., Peptide sequencing: the utility of chemical ionization mass spectrometry, *Biomed. Mass Spectrom.*, 4, 159, 1977.

121. Hunt, D. F., Buko, A. M., Ballard, J. M., Shabanowitz, J., and Giardini, A. B., Sequence analysis of polypeptides by collision activated dissociation on a triple quadrupole mass spectrometer, *Biomed. Mass Spectrom.*, 8, 397, 1981.

122. Bradley, C. V., Howe, I., and Beynon, J. H., Analysis of underivatized peptide mixtures by collision-induced dissociation of negative ions, *J. Chem. Soc. Chem. Commun.*, 562, 1980.

123. Bradley, C. V., Howe, I., and Beynon, J. H., Sequence analysis of underivatized peptides by negative ion chemical ionization and collision-induced dissociation, *Biomed. Mass Spectrom.*, 8, 85, 1981.

124. Hunt, D. F., Stafford, G. C., Crow, F. W., and Russell, J. W., Pulsed positive negative ion chemical ionization mass spectrometry, *Anal. Chem.*, 48, 2098, 1976.

125. Michnowicz, J. and Munson, B., Studies in chemical ionization mass spectrometry: 17-hydroxy steroids, *Org. Mass Spectrom.*, 6, 765, 1972.

126. Lin, Y. Y. and Smith, L. L., Recognition of functional groups by chemical ionization mass spectrometry, *Biomed. Mass Spectrom.*, 5, 604, 1978.

127. Dzidic, I. and McCloskey, J. A., Chemical ionization mass spectrometry using ammonia reagent gas. Selective protonation of conjugated ketones, *Org. Mass Spectrom.*, 6, 939, 1972.

128. Roy, T. A., Field, F. H., Lin, Y. Y., and Smith, L. L., Hydroxyl ion negative chemical ionization mass spectra of steroids, *Anal. Chem.*, 51, 272, 1979.

129. Radford, T. and DeJongh, D. C., Carbohydrates, in *Biochemical Applications of Mass Spectrometry*, Waller, G. R., Ed., Interscience, New York, 1972, chap. 12.

130. Radford, T. and DeJongh, D. C., Carbohydrates, in *Biochemical Applications of Mass Spectrometry*, 1st Suppl. Vol., Waller, G. R. and Dermer, O. C., Eds., Interscience, New York, 1980, chap. 12.

131. Fales, H. M., Milne, G. W. A., and Vestal, M. L., Chemical ionization mass spectrometry of complex molecules, *J. Am. Chem. Soc.*, 91, 3682, 1969.

132. Dougherty, R. C., Horton, D., Philips, K. D., and Wander, J. D., The high resolution mass spectrum of 2-acetamido-1,3,4,6-tetra-O-acetyl-2-dioxy-α-D-glucopyranose, *Org. Mass Spectrom.*, 7, 805, 1973.

133. Dougherty, R. C., Roberts, J. D., Binkley, W. W., Chizhov, O. S., Kadentsev, V. I., and Solov'yov, A. A., Ammonia-isobutane chemical ionization mass spectra of oligosaccharide peracetates, *J. Org. Chem.*, 39, 451, 1974.

134. Chizhov, O. S., Kadentsev, V. I., Solov'yov, A. A., Brinkley, W. W., Roberts, J. D., and Dougherty, R. C., Oligosaccharide acetate mass spectra obtained by addition of ammonium ions (transl.), *Dokl. Akad. Nauk. S.S.S.R.*, 217, 511, 1974.

135. Chizhov, O. S., Kadentsev, V. I., Solov'yov, A. A., Levonowich, P. F., and Dougherty, R. C., Polysaccharide sequencing by mass spectrometry. Chemical ionization spectra of permethylglycosyl-lalditols, *J. Org. Chem.*, 41, 3425, 1976.

136. Ganguly, A. K., Cappuccino, N. F., Fujiwara, H., and Bose, A. K., Convenient mass spectral technique for structural studies in oligosaccharides, *J. Chem. Soc. Chem. Commun.*, 148, 1979.

137. Heaney Kieras, J., Kieras, F., and Bowen, D., 2-O-methyl-D-glucoronic acid, a new hexuronic acid of biological origin, *Biochem. J.*, 155, 181, 1976.

138. Lyle, M., Palante, S., Head, K., and Fenselau, C., Synthesis and characterization of glucoronides of cannabinol, cannabidiol, Δ⁹-tetrahydrocannabinol, and Δ⁸-tetrahydrocannabinol, *Biomed. Mass Spectrom.*, 4, 190, 1977.

139. Johnson, L. P., Subba Rao, S. C., and Fenselau, C., Pyridine as a reagent gas for the characterization of glucoronides by chemical ionization mass spectrometry, *Anal. Chem.*, 50, 2022, 1978.

140. Fenselau, C., Cotter, R., and Johnson, L., Mass spectral techniques for the analysis of glucoronides, *Adv. Mass Spectrom.*, 8, 1159, 1980.

141. Bruins, A. P., Negative ion desorption chemical ionization mass spectrometry of some underivatized glucoronides, *Biomed. Mass Spectrom.*, 8, 31, 1981.

142. Wilson, M. S., Dzidic, I., and McCloskey, J. A., Chemical ionization mass spectrometry of nucleosides, *Biochim. Biophys. Acta*, 240, 623, 1971.

143. McCloskey, J. A., Futrell, J. H., Elwood, T. A., Schram, K. H., Panzica, R. P., and Townsend, L. B., Determination of relative glycosyl bond strengths in nucleosides by chemical ionization mass spectrometry. A comparative study of 7- and 9-β-D-ribofuranosylpurines, *J. Am. Chem. Soc.*, 95, 5762, 1973.

144. Wilson, M. S. and McCloskey, J. A., Chemical ionization mass spectra of nucleosides: mechanism of ion formation and estimation of proton affinity, *J. Am. Chem. Soc.*, 97, 3436, 1975.

145. Esmins, E. L., Freyne, E. J., Vanbroeckhoven, J. H., and Alderwiereldt, F. C., Chemical ionization desorption spectrometry as an additional tool for the structure elucidation of nucleosides, *Biomed. Mass Spectrom.*, 7, 377, 1980.

146. Hunt, D. F., Shabanowitz, J., Botz, F. K., and Brent, D. A., Chemical ionization mass spectrometry of salts and thermally labile organics with field desorption emitters as solid probes, *Anal. Chem.*, 49, 1160, 1977.

Chapter 6

SELECTED TOPICS IN CHEMICAL IONIZATION MASS SPECTROMETRY

I. INTRODUCTION

This chapter reviews four specific topics, isotope exchange reactions in CI systems, stereochemical effects in chemical ionization mass spectra, the collision-induced dissociation of ions produced by CI, and reactant ion monitoring. These topics are not covered in detail in earlier chapters.

II. ISOTOPE EXCHANGE REACTIONS IN CHEMICAL IONIZATION STUDIES

The exchange of hydrogen for deuterium in organic molecules has found wide use in structural studies by mass spectrometry.[1,2] Normally the exchange is carried out in solution reactions before the sample is introduced into the mass spectrometer, but in some cases the exchange reaction can be carried out in the inlet system of the mass spectrometer. There now is substantial evidence that at the pressures applicable in chemical ionization, exchange of hydrogen for deuterium occurs in the ion source of the mass spectrometer. Hunt et al.[3] have shown that the hydrogens bonded to heteroatoms in alcohols, phenols, carboxylic acids, amines, amides, and thiols undergo rapid exchange for deuterium under proton-transfer CI conditions using D_2O as reagent gas. The mass shift on substituting D_2O for H_2O as reagent gas thus provides a count of the number of such hydrogens in the molecule. In a similar fashion the use of NH_3[4] or CH_3OD[5] as reagent gas permits the differentiation of primary, secondary, and tertiary amines from the mass shift when the deuterated reagent gas is substituted for the undeuterated reagent gas.

Although it was reported[3] that aromatic hydrogens do not exchange when D_2O is used as the reagent gas in proton transfer CI, this report is not true. The gas phase exchange of aromatic hydrogens was observed first by Beauchamp and co-workers.[6] Using ICR techniques they observed sequential replacement of hydrogen by deuterium when protonated benzene reacted with D_2O. A number of fluorine- and methyl-substituted benzenes also were observed to exchange with D_2O although the rate and extent of exchange was found to be very structure-dependent. Thus, although o- and p-xylene exchanged all aromatic hydrogens rapidly, m-xylene showed exchange of one hydrogen only. No exchange was observed for benzenes containing strong electron-releasing or electron-donating substituents. Many of these protonate on the substituent and it was concluded that protonation of the aromatic ring was a prerequisite for exchange to occur. Martinson and Buttrill[7] recorded similar observations and reached similar conclusions in a study of the positive ion D_2O CI mass spectra of a variety of aromatic compounds.

In studies involving negative ions Stewart et al.[8] have shown that the [M−H]⁻ ions from simple esters, olefins, acetylenes, allenes, and toluene undergo exchange when allowed to react with D_2O. These experiments were performed in a flowing afterglow apparatus. In a later study[9] the [M−H]⁻ ions from ketones and aldehydes were shown to exchange with CH_3OD under flowing afterglow conditions.

An extensive study of the exchange of hydrogen for deuterium in a variety of compounds using both Brønsted acid positive ion CI and Brønsted base negative ion CI has been reported by Hunt and Sethi.[10] In positive ion studies using D_2O as reagent gas they observed that the dominant ion in the CI mass spectra of benzene, toluene, *o*-xylene, and *p*-xylene corresponded to the MD⁺ ion in which all aromatic hydrogens had been replaced by deuterium. Slow exchange of hydrogens was observed for *m*-xylene and mesitylene while no exchange of aromatic hydrogens was observed for aromatic compounds containing amino, hydroxy, alkoxy, acetyl and nitrile substituents, for polyaromatic compounds, and for the metallocenes investigated. Exchange of hydrogens bonded to heteroatoms did, of course, occur.

When the more basic reagent gas C_2H_5OD was used the rate of exchange of the aromatic hydrogens in *m*-xylene and mesitylene increased substantially, the most prominent ion corresponding to exchange of all aromatic hydrogens. The exchange rate for the oxygen-containing derivatives also increased although there was some evidence that those hydrogens *ortho* or *para* to the substituent were exchanged more rapidly. In addition, exchange of hydrogens in polyaromatic compounds, except pyrene, also occurs with C_2H_5OD as reagent gas. Replacement of aromatic hydrogens was not observed for aminobenzenes, benzonitrile, acetophenone, or ferrocene.

When the even more basic reagent ND_3 was used slow exchange of aromatic hydrogens in *m*-toluidine and *m*-phenylenediamine, ferrocene, and vinylferrocene was observed. Aniline, *o*- and *p*-toluidine, *o*- and *p*-phenylenediamine, ruthenocene, hydroxymethylferrocene, and ferrocene carboxaldehyde did not exchange aromatic hydrogens with any of the reagent gases.

Negative ion studies were carried out by Hunt and Sethi[10] using OD⁻ (in D_2O), $C_2H_5O^-$ (in C_2H_5OD), and ND_2^- (in ND_3). When OD⁻ was used as reagent ion the [M−H]⁻ ion of alkyl benzenes showed exchange of all benzylic hydrogens. Slow exchange was observed for hydrogen atoms adjacent to the carbonyl group of simple ketones. The weaker base $C_2H_5O^-$ did not abstract a proton from alkyl benzenes but did promote the exchange reaction in simple ketones; the most prominent ion corresponded to [M−H]⁻ with all enolizable hydrogens replaced by deuterium. Incomplete exchange of enolizable hydrogens was observed for conjugated carbonyl compounds.

The strong base NH_2^- abstracts on aromatic proton from a variety of compounds with the resulting [M−H]⁻ ion undergoing rapid exchange of aromatic hydrogens for benzene, naphthalene, anthracene, and phenanthrene, with slower exchange for pyrene, chrysene, and benz[*a*]pyrene. The rate of exchange also was slow for toluene. Exchange of the aromatic hydrogens in the organometallic compounds ferrocene, ferrocene carboxaldehyde, phenyl ferrocene, and (toluene) chromium tricarbonyl also was observed.

The exchange reaction is believed to be the result of collisions between sample ions (HR⁻, H_2RD^+) and reagent gas neutrals, R′OD. At 0.5 to 1.0 torr pressure each ion suffers up to several hundred collisions with neutrals (mostly reagent gas) before exiting from the mass spectrometer ion source. By contrast, multiple collisions between a given sample molecule and the reagent ion population are highly improbable due to the low concentration of each. This precludes significant interchange by the sequential Reactions 1 to 3. Rather, the incorporation of deuterium into the sample ion must occur during the lifetime of the ion/molecule complex formed when the sample ion and the deuterium-labeled reagent gas collide.

$$H_2R + R'OD_2^+ \rightarrow H_2RD^+ + R'OD \tag{1}$$

$$H_2RD^+ + R'OD \rightarrow HRD + R'ODH^+ \tag{2}$$

$$HRD + R'OD_2^+ \rightarrow HRD_2^+ + R'OD \tag{3}$$

There is substantial evidence[11-13] that the potential energy profile for proton transfer between an ion and a neutral contains two minima with an intermediate potential energy barrier, as illustrated schematically in Figure 1 for the case of negative ions. The two minima correspond to R′O⁻ hydrogen bonded to HRH and RH⁻ hydrogen bonded to R′OH. The exchange reaction occurs when the ion HR⁻ reacts with the reagent gas R′OD and the system has sufficient energy to overcome the intermediate energy barrier (lower curve, Figure 1) to form HR-D---ŌR′ with an energy part way up the curve leading to HRD + ŌR′. At this point the hydrogen bonding is weak, permitting rotation of HRD leading to DR-H---ŌR′, which exits to DR⁻ + R′OH as summarized in Scheme 1.

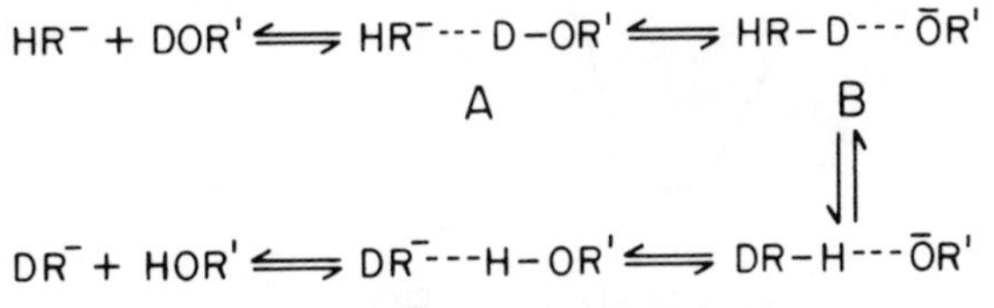

$$HR^- + DOR' \rightleftharpoons HR^- \text{---} D-OR' \rightleftharpoons HR-D \text{---} \bar{O}R'$$

$$\qquad\qquad\qquad\qquad\qquad A \qquad\qquad\qquad\qquad B$$

$$\qquad\qquad\qquad\qquad\qquad\qquad\qquad\qquad\qquad\qquad \Updownarrow$$

$$DR^- + HOR' \rightleftharpoons DR^- \text{---} H-OR' \rightleftharpoons DR-H \text{---} \bar{O}R'$$

SCHEME 1

A similar type of potential energy profile undoubtedly applies for proton transfer from a Brønsted acid to a neutral molecule and a similar mechanism, leads to H/D exchange as indicated in Scheme 2.

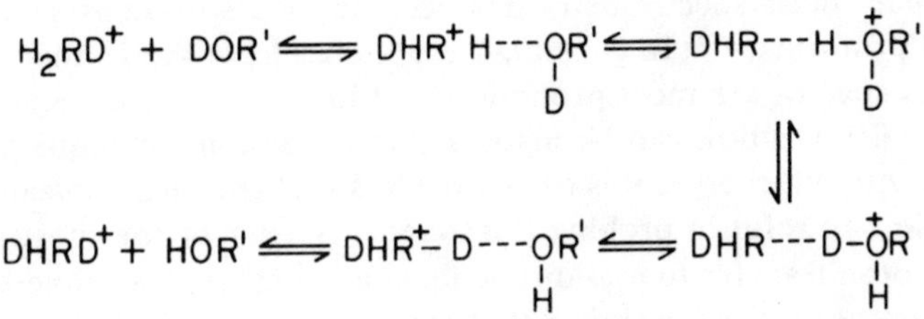

$$H_2RD^+ + DOR' \rightleftharpoons \underset{D}{DHR^+H} \text{---} OR' \rightleftharpoons DHR \text{---} H - \underset{D}{\overset{+}{O}R'}$$

$$\qquad\qquad\qquad\qquad\qquad\qquad\qquad\qquad\qquad\qquad\qquad\qquad \Updownarrow$$

$$DHRD^+ + HOR' \rightleftharpoons \underset{H}{DHR^+} D \text{---} OR' \rightleftharpoons DHR \text{---} D - \underset{H}{\overset{+}{O}R}$$

SCHEME 2

The potential energy profiles in Figure 1 also explain why different reagent systems exhibit different abilities to promote exchange. As shown, if the initial proton transfer reaction is highly exothermic the intermediate energy barrier may be sufficiently high as to prevent solvation of the sample ion reaching the intermediate B in which exchange of hydrogen for deuterium occurs. In general H/D exchange will be more efficient when the initial chemical ionization reaction is only slightly exothermic. Thus, in positive ion studies, exchange is more efficient when $C_2H_5OD(PA[C_2H_5OH] = 190$ kcal mol^{-1}) is the reagent gas than when D_2O $(PA[H_2O] = 173$ kcal mol^{-1}) is the reagent gas. Similarly, in negative ion studies, exchange is more efficient when C_2H_5OD $(PA[C_2H_5O^-] = 376$ kcal mol^{-1}) is used than when D_2O $(PA[OH^-] = 391$ kcal mol^{-1}) is used.

The exchange of hydrogen for deuterium has not been extensively employed in structural studies by chemical ionization although it appears possible to carry out this exchange in the ion source. It is relatively easy to establish the number of hydrogens attached to heteroatoms by isotope exchange. In favorable cases it may be possible to establish the number of enolizable hydrogens, the number of benzylic hydrogens, or the number of aromatic hydrogens using the methodology outlined above.

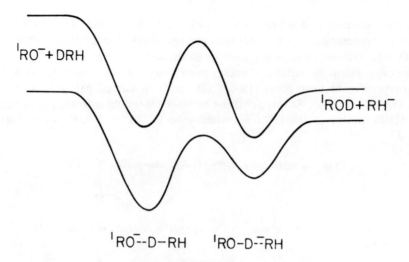

FIGURE 1.   Schematic potential energy profiles for reaction of RO^- with DRH.

III. STEREOCHEMICAL EFFECTS ON CHEMICAL IONIZATION MASS SPECTRA

Electron impact mass spectrometry has been applied with considerable success to stereochemical problems.[14-16] In general, these studies have shown that stereochemical effects on mass spectra are most pronounced at low ion energies. Since, in principle at least, chemical ionization can be made a soft ionization technique producing ions of low average internal energy, it is to be expected that chemical ionization mass spectrometry should be useful in probing the stereochemistry of the substrate molecules.

Studies of proton transfer to α,ω-difunctional alkanes, such as dimethoxyalkanes,[17] diols,[18] and diamines,[19,20] have shown that the proton is particularly strongly bonded to these difunctional molecules as the result of intramolecular hydrogen bonding; the internal solvation results in an enhanced proton affinity of the difunctional molecule compared to the corresponding monofunctional molecule. A considerable number of CI studies have used the presence or absence of internal solvation to probe the stereochemistry of the substrate molecule. The occurrence of internal solvation may be signalled by an enhanced stability of the MH^+ ion for exothermic protonation reactions or by an enhanced proportion of simple proton transfer vs. alternative reaction channels for nearly thermoneutral protonation reactions.

In early work Longevialle et al.[21] studied the $i\text{-}C_4H_{10}$ CI mass spectra of a series of steroidal 1,2- and 1,3-aminoalcohols. They observed only MH^+ when the configuration of the aminoalcohol permitted intramolecular hydrogen bonding. When such hydrogen bonding was not permitted by the molecular stereochemistry, appreciable fragmentation of MH^+ by elimination of H_2O was observed. The results obtained correlated well with the evidence for hydrogen bonding between the amine and hydroxyl functions in the neutral molecules as deduced from the IR spectra. In an extension of this work Longevialle and co-workers[22] examined the $i\text{-}C_4H_{10}$ CI mass spectra of a series of conformationally stable β-aminoalcohols and recorded the % elimination of H_2O from MH^+ as a function of the dihedral angle between the two functional groups. For angles <90° no loss of H_2O was observed; as the angle increased beyond 90° the loss of H_2O increased, at 180° reaching $\sim$50% for the aminoalcohols and $\sim$60% for the N,N-dimethylaminoalcohols. These values were taken as the respective percentages of protonation at the hydroxyl function. In subsequent work[23] they used the percentage of H_2O

Table 1
PARTIAL i-C₄H₁₀ CI MASS SPECTRA OF 1,4-CYCLOHEXANEDIOLS

	Relative intensity	
	cis	*trans*
M·C₄H₉⁺	12	42
MH⁺	100	58
[MH⁺−H₂O]	12	100
[MH⁺−2H₂O]	19	39

loss from MH⁺ to evaluate the conformer populations of conformationally mobile β-aminoalcohols (and N,N-dimethyl derivatives) with results in good agreement with those obtained from IR and NMR studies. They assumed that in conformations where internal solvation was possible no loss of H₂O from MH⁺ would occur, while in conformations where internal solvation was not possible 50% (aminoalcohols) or 60% (N,N-dimethylaminoalcohols) of the MH⁺ ions would eliminate H₂O. They further assumed that loss of H₂O from the hydroxyl-protonated form in the nonsolvating conformation was rapid with respect to conformational change.

Winkler and McLafferty[24] observed that the proton transfer (CH₄ and i-C₄H₁₀) CI mass spectra of stereoisomeric cycloalkane diols showed substantial differences characteristic of their stereochemistry. For the *cis* isomers of 1,3- and 1,4-cyclohexanediol formation of an intramolecular hydrogen bond resulted in much higher abundances of MH⁺ compared to the *trans* isomers. As an example Table 1 records partial i-C₄H₁₀ CI mass spectra of the 1,4-cyclohexanediols.[25] (The spectra disagree quantitatively with those reported by Winkler and McLafferty although the same conclusions result.) The stability of the proton bridge in *cis*-1,3-cyclohexanediol is decreased by a sterically interfering 5-methyl group and enhanced by an interfering 5-hydroxy group.[24] The proton transfer CI mass spectra of *cis*- and *trans*-1,2-cyclohexanediol are essentially identical;[24,25] apparently the flexibility of the cyclohexane ring is sufficient to make the distance between the two hydroxyl functions the same for the two isomers. In the more rigid 1,2-cyclopentanediols the *cis* isomer shows a much more abundant MH⁺ ion than the *trans* isomer.[24] When the conformational flexibility of the 1,2-cyclohexanediols is reduced by forming the trimethylsilyl derivatives, significant differences are observed[26] in the i-C₄H₁₀ CI mass spectra, with the [MH⁺−(CH₃)₃SiOH] ion being much more abundant for the *trans* isomer than for the *cis* isomer. Similarly the i-C₄H₁₀ CI mass spectra of the trimethylsilyl ethers of stereoisomeric 3,4-dimethyl-1,2-cyclopentanediols show substantial differences. The *cis*-diol ethers show abundant MH⁺, [MH⁺−(CH₃)₃SiOH], and [MH⁺−2(CH₃)₃SiOH] ions while the *trans*-diol ethers show no MH⁺, the major ion (>95% of ionization) observed being [MH⁺−2(CH₃)₃SiOH].[26] Minor differences in the *cis*-diol ether spectra can be correlated with the orientation of the methyl groups. The i-C₄H₁₀ CI mass spectra of the trimethylsilyl ethers of the 2,3-dimethyl-1,4-cyclopentanediols show abundant MH⁺ ion signals for the *cis*-diol isomers with no MH⁺ ion signals for the *trans*-diol isomers.[27]

The i-C₄H₁₀ CI mass spectra[28] of the methyl ethers of the 2-methyl-1,3-cyclopentanediols and 2-methyl-1,3-cyclohexanediols are presented in Table 2. The *trans* ethers show essentially no MH⁺ ion signal in contrast to the abundant MH⁺ ion signals observed for the *cis* ethers. When the C₂-methyl is also *cis* the proton bridge is destabilized and significantly lower MH⁺ ion signals are observed. The MH⁺ ion is abundant in the i-C₄H₁₀ CI mass spectrum of 2,5-*diendo*-protoadamantanediol due to internal solva-

Table 2
i-C$_4$H$_{10}$ CI MASS SPECTRA OF CYCLOPENTANEDIOL AND CYCLOHEXANEDIOL METHYL ETHERS

	MH$^+$	[M−H]$^+$	[MH$^+$−CH$_3$OH]	[MH$^+$−2CH$_3$OH]	[MH$^+$−46]
			% Total ionization		
(cyclopentane OCH$_3$/CH$_3$/OCH$_3$)	0	0.8	94.8	3.0	1.5
(cyclopentane OCH$_3$/CH$_3$/OCH$_3$)	74.9	0.1	17.5	6.8	0.8
(cyclopentane OCH$_3$/CH$_3$/OCH$_3$)	59.8	0.3	33.3	5.4	1.2
(cyclohexane OCH$_3$/CH$_3$/OCH$_3$)	0.7	1.2	38.2	34.1	25.9
(cyclohexane OCH$_3$/CH$_3$/OCH$_3$)	68.8	0.1	19.8	8.6	2.8
(cyclohexane OCH$_3$/CH$_3$/OCH$_3$)	51.5	0.3	19.6	16.5	12.2

tion of the proton but is of very low abundance in the *diexo* and the two *exo-endo* epimers.[29] Minor differences in fragment ion abundances for the last three epimers can be correlated with structure.

The proton transfer CI of the diacetate derivatives of cyclic diols have been studied extensively. The CH$_4$ and i-C$_4$H$_{10}$ CI mass spectra[30] of the 1,2-cyclopentanediol and 1,2-cyclohexanediol diacetates are summarized in Table 3. The i-C$_4$H$_{10}$ CI mass spectra are characterized by a more abundant MH$^+$ ion for the *cis* isomer in each case. The CH$_4$ CI mass spectra are characterized by a more abundant [MH$^+$−HOAc] fragment ion and less fragmentation of this species by further loss of HOAc or C$_2$H$_2$O for the *trans* isomers. Similar results have been reported for the proton transfer CI of the stereoisomeric diacetates of 1,3-cyclopentanediol and 1,3-cyclohexanediol,[30] 3,4-di-methyl-1,2-cyclopentanediol,[26] and 2-methyl-1,3-cyclopentanediol.[28] The lower MH$^+$ ion abundances and the unusual stability of the [MH$^+$−HOAc] species for the *trans* epimers have been attributed to anchimeric assistance (Reaction 4) leading to dioxolan-ylium and dioxanylium ions.

$$\text{(reaction scheme)} \longrightarrow \text{(dioxolanylium ion)} \quad + \quad \text{HOAc} \tag{4}$$

In related studies it has been shown[31] that *cis* 3-methoxy and 3-trimethylsiloxy cyclo-pentyl and cyclohexyl acetic acid esters show abundant MH⁺ ion in their i-C_4H_{10} CI mass spectra while the *trans* isomers do not. Only very small differences were observed in the CH_4 CI mass spectra. Similarly *cis*-4-methoxycyclohexane carboxylic acid ethyl ester shows a much more intense MH⁺ ion in the i-C_4H_{10} CI mass spectrum than does the *trans* isomer.[32] In the same study the i-C_4H_{10} CI mass spectra of the stereoisomeric 2,6-dimethyl-4-hydroxytetrahydropyranes, 2,4,6-trimethyl-4-hydroxytetrahydropy-ranes, and 2,7-dimethoxy-*cis*-decalins were determined; the MH⁺ ion was found to be particularly abundant when a proton bridge between the two oxygen functions was permitted by the stereochemistry of the molecule.

When the stereochemistry of the molecule permits intramolecular hydrogen bonding the proton affinity is increased relative to the epimer where such bonding cannot occur. Thus, with a suitable choice of protonating agent, it should be possible to selectively protonate the epimer of higher-proton affinity. The NH_4^+ ion is a weak protonating agent (PA[NH_3] = 205 kcal mol⁻¹) and is known[33] to react by addition when protona-tion is not possible. Winkler and Stahl[34] have shown that the NH_3CI mass spectra of cyclic diols reflect the stereochemistry quite strongly. Their results for the epimeric 1,3- and 1,4-cyclohexanediols and 1,2-cyclopentanediols are presented in Table 4. For the *cis* cyclohexanediols abundant MH⁺ ions are observed; however for the *trans* iso-mers practically no MH⁺ ions are observed, the NH_4^+ ion reacting much more exten-sively to form [M + NH_4]⁺. A similar result can be seen (Table 1) in the i-C_4H_{10} CI of the 1,4-cyclohexanediols where the MH⁺/[M + C_4H_9]⁺ ratio is much greater for the *cis* isomer than for the *trans* isomer. The two 1,2-cyclopentanediols show predominant formation of [M + NH_4]⁺. This has been attributed[34] to chelation of the NH_4^+ ion (Struc-ture A) by the 1,2-diol, which is possible for both epimers.

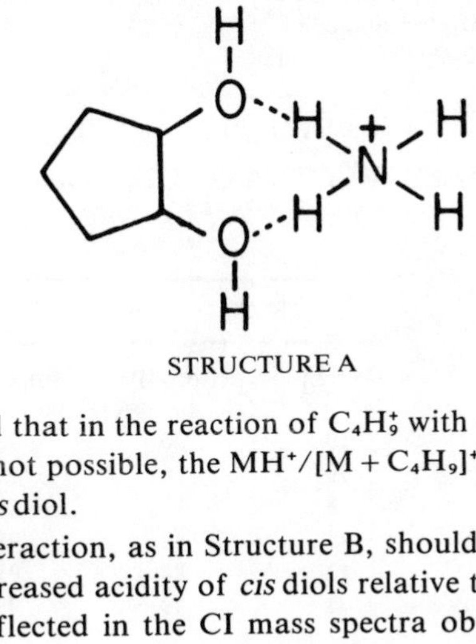

STRUCTURE A

Winkler and Stahl noted that in the reaction of $C_4H_9^+$ with the 1,2-cyclopentanediols,[24] where such chelation is not possible, the MH⁺/[M + C_4H_9]⁺ ratio is much larger for the *cis* diol than for the *trans* diol.

Functional group interaction, as in Structure B, should stabilize the [M−H]⁻ ion of diols and lead to an increased acidity of *cis* diols relative to *trans* diols; this difference in acidity should be reflected in the CI mass spectra obtained using weak Brønsted base reactants.

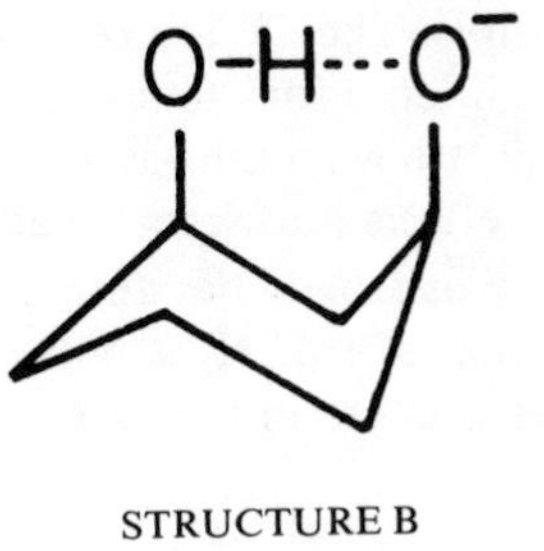

STRUCTURE B

Table 3
CI MASS SPECTRA OF 1,2-CYCLOPENTANEDIOL AND 1,2-CYCLOHEXANEDIOL ACETATES[a]

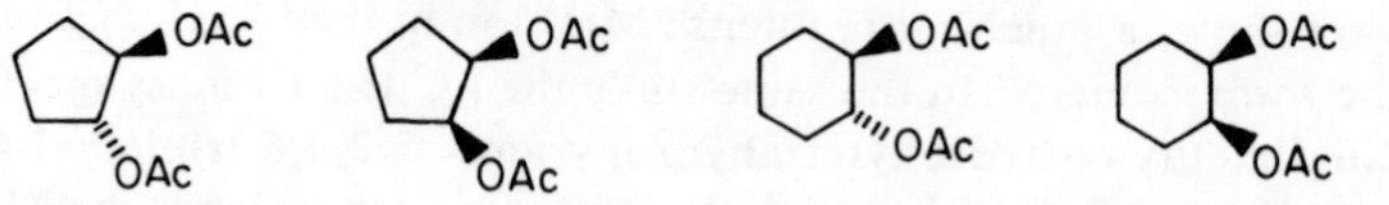

	CH_4	i-C_4H_{10}	CH_4	i-C_4H_{10}	CH_4	i-C_4H_{10}	CH_4	i-C_4H_{10}
MH^+	1.9	32.1	1.7	45.2	2.1	33.8	1.4	46.5
$[MH^+-HOAc]$	75.6	56.9	21.1	36.2	59.5	52.6	30.2	27.5
$[MH^+-2HOAc]$	6.6	0.4	16.9	4.5	15.3	1.0	14.8	2.3
$[MH^+-HOAc-C_2H_2O]$	4.3	0.2	38.3	3.7	7.7	—	42.4	9.3

[a] Intensities as % of total ionization.

Table 4
NH_3 CI MASS SPECTRA OF CYCLOALKANEDIOLS

	% Total ionization	
	$[M+NH_4]^+$	MH^+
cis-1,3-Cyclohexanediol	29	51
trans-1,3-Cyclohexanediol	54	2.2
cis-1,4-Cyclohexanediol	15	64
trans-1,4-Cyclohexanediol	19	0.5
cis-1,2-Cyclopentanediol	53	0.4
trans-1,2-Cyclopentanediol	56	0.4

Table 5
F^- AND Cl^- CI MASS SPECTRA OF CYCLOALKANEDIOLS

	% Total ionization			
	F^- CI		Cl^- CI	
	$[M+F]^-$	$[M-H]^-$	$[M+Cl]^-$	$[M-H]^-$
cis-1,3-Cyclohexanediol	10	90	95	4.6
trans-1,3-Cyclohexanediol	39	36	99	1.1
cis-1,4-Cyclohexanediol	8	92	88	12
trans-1,4-Cyclohexanediol	45	44	98	1.7
cis-1,2-Cyclopentanediol	12	88	95	4.7
trans-1,2-Cyclopentanediol	42	39	99	1.1

Table 5 summarizes the results obtained by Winkler and Stahl[34] for the F^- and Cl^- CI of a number of epimeric cyclic diols. The Cl^- ion reacts predominantly by attachment, however the $[M-H]^-$ ion is more intense for the *cis* diols. This difference is much more pronounced in the F^- CI mass spectra where the abundance of $[M-H]^-$ for the *cis* diols relative to the *trans* diols clearly reflects the increased acidity of the *cis* epimers.

The stabilization of $[M-H]^-$ by hydrogen bonding in the *cis* diols also is reflected in the smaller extent of fragmentation of $[M-H]^-$ (by elimination of H_2 or H_2O) for the *cis* epimers relative to the *trans* epimers in the OH^- CI mass spectra of the diols of Table 5.[35]

Table 6
CH₄ CI MASS SPECTRA
OF DICARBOXYLIC
ACID DERIVATIVES[a]

Structure	MH⁺	[MH⁺−ROH]
maleic acid (HC=CH, each C–OH with =O)	10	100
HC–COOH / HOOC–C–H (fumaric)	100	14
H₃C ⌐COOH ⌐COOH	16	100
HOOC⌐ ⌐CH₃ ⌐COOH	100	24
⌐COOEt ⌐COOEt	13	100
EtOOC⌐ ⌐COOEt	100	10[b]
H⌐ H⌐ ⌐COOφ ⌐COOφ	8	100
φOOC⌐ ⌐COOφ	100	100

a Intensities as % of base peak.
b [MH⁺−C₂H₄] = 25.

In contrast to the compounds discussed above, where interaction of the two functional groups stabilizes the MH⁺ ion, interaction of the two carboxylate groups in dicarboxylic acid derivatives leads to a more facile fragmentation of MH⁺.[36,37] As a result dicarboxylic acid derivatives which have a *cis* configuration about a double bond show more extensive fragmentation than those which have a *trans* configuration. Results obtained[37] in the CH₄ CI of maleic and fumaric acids and derivatives are summarized in Table 6, where it is seen that the effect is quite dramatic. The thermochemically favored site of protonation is the carbonyl oxygen.[38] For monocarboxylic compounds or *trans* dicarboxylic compounds elimination of ROH requires (Reaction 5) a symmetry-forbidden 1,3-H migration, a reaction with a high energy barrier.[39] By contrast, interaction of two carboxylate functions provides (Reaction 6) an alternative mode of H migration which is not symmetry-forbidden. In addition, the elimination of ROH is favored by formation of the stable cationated anhydride structure.

(5)

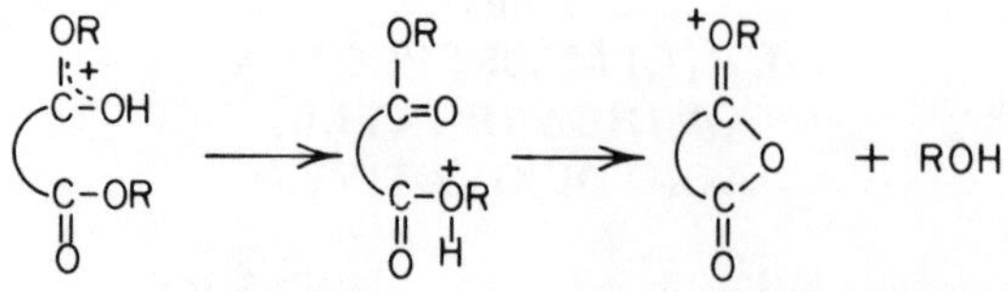

$$\begin{array}{c} \text{OR} \\ \overset{||}{\underset{||}{\text{C}}\text{=}}\text{OH} \\ \overset{\text{C-OR}}{\underset{||}{\text{C}}} \\ \text{O} \end{array} \longrightarrow \begin{array}{c} \text{OR} \\ \overset{|}{\text{C=O}} \\ \overset{+}{\text{C-OR}} \\ \overset{||}{\text{O}}\overset{}{\text{H}} \end{array} \longrightarrow \begin{array}{c} \overset{+}{\text{OR}} \\ \overset{||}{\text{C}}\diagdown \\ \overset{}{\text{C}}\diagup\text{O} \\ \overset{||}{\text{O}} \end{array} + \text{ROH}$$

(6)

In other studies of stereochemical effects on CI mass spectra Michnowicz and Munson[40] observed that 5α-androstane-3-one-17β-ol and 5β-androstane-3-one-17β-ol were distinguishable from their H_2 CI mass spectra with the former showing the relative intensities $MH^+/[MH^+-H_2O]/[MH^+-2H_2O] = 4/2.5/1$ compared to 1/3/3 for the latter. In later work[41] similar differences were observed in the H_2 CI mass spectra of the epimeric pairs: 5α-androstane-3,17-dione and 5β-androstane-3,17-dione; 5α-cholestane-3-one and 5β-cholestane-3-one. For each pair the β isomer showed more extensive fragmentation by loss of H_2O from MH^+. The results of the two studies can be rationalized by noting that the *cis* form of the A/B ring junction in the 5β compounds allows a relatively close interaction between the 3-keto function and the hydrogens on the B ring.

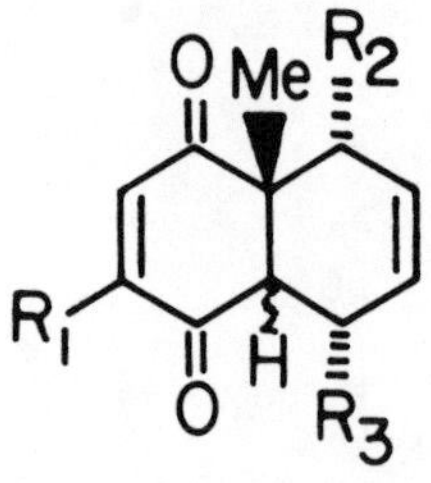

STRUCTURE C 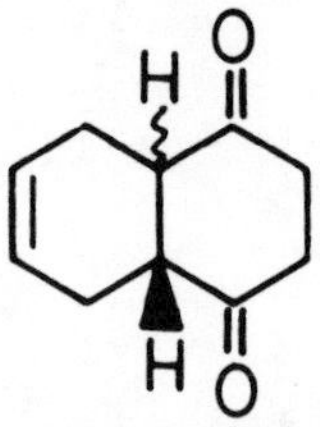   STRUCTURE D 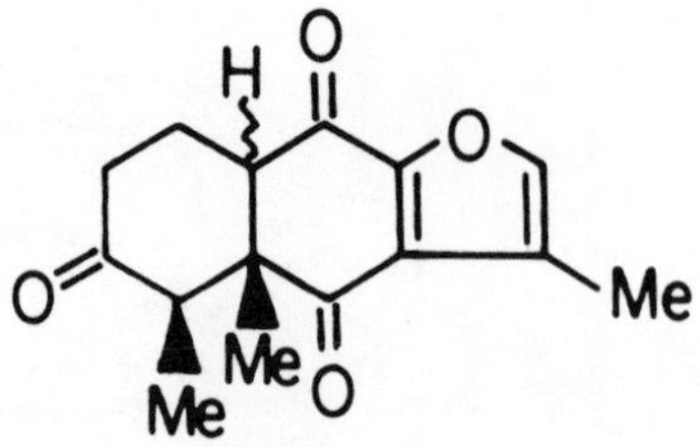   STRUCTURE E

Schwarz and co-workers[42] have observed that the retro-Diels-Alder fragmentation of protonated or ethylated Structures C (several $R_1, R_2,$ and R_3 substituents), D and E are highly stereospecific to the *cis* ring-fused epimers. The retro-Diels-Alder reaction can be written as Reaction 7 for Structure C with similar formulations for Structures D and E. By contrast, elimination of H_2O from the protonated forms of Structure C was more prominent for the *trans* ring-fused epimers.

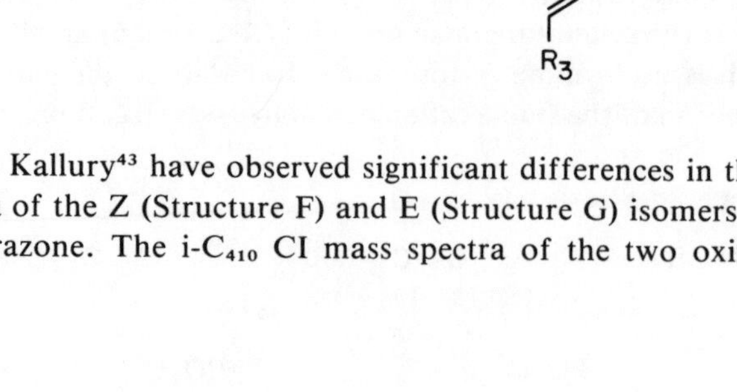

(7)

Harrison and Kallury[43] have observed significant differences in the proton transfer CI mass spectra of the Z (Structure F) and E (Structure G) isomers of benzoin oxime and phenylhydrazone. The i-C_{410} CI mass spectra of the two oximes are shown in Figure 2.

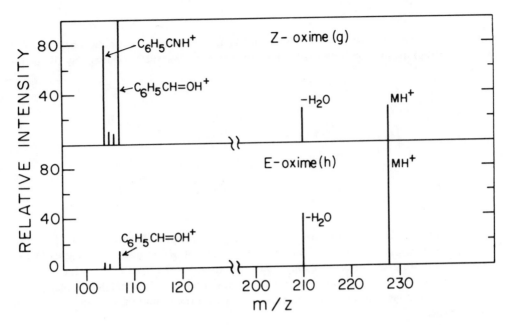

FIGURE 2. i-C$_4$H$_{10}$ CI mass spectra of Z- and E-benzoin oximes. Data from Reference 43.

Y = OH, NHC$_6$H$_5$

STRUCTURE F   STRUCTURE G

For both the oximes and the phenylhydrazones the E isomer gave a more abundant MH$^+$ ion signal with less extensive fragmentation. The increased stability of MH$^+$ for the E isomers was attributed to more effective internal solvation of the added proton by the benzoin hydroxyl and the doubly bonded nitrogen.

Budzikiewicz and Busker[44] have observed that the [M + C$_4$H$_9$]$^+$/[M−H]$^+$ ratio observed in the i-C$_4$H$_{10}$ CI of Z-octadecenes is ∼2 compared to a value <1 for the E isomers. This has been attributed to less shielding of the π system from the approaching C$_4$H$_9^+$ for the Z isomers. Hydride ion abstraction from various CH$_2$ groups would not be expected to be affected by the double bond configuration.

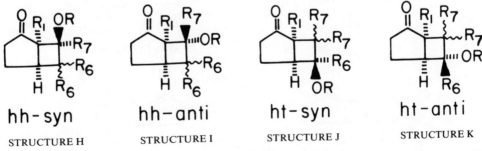

hh−syn   hh−anti   ht−syn   ht−anti

STRUCTURE H   STRUCTURE I   STRUCTURE J   STRUCTURE K

The regio- and stereoisomeric bicyclo[3,2,0]-heptanone-2 derivatives (Structures H to K) have been extensively investigated by i-C$_4$H$_{10}$ CI mass spectrometry for several substituted series and for $R = \overset{\text{O}}{\overset{\|}{C}}CH_3$ or $Si(CH_3)_3$.[45-47] A ready distinction between the hh and ht regioisomers can be made. The hh isomers show facile cleavage by the processes

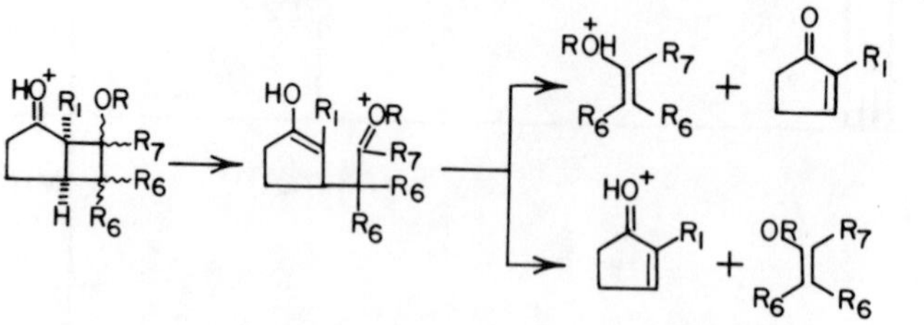

(8)

while the ht regioisomers show little cleavage by these routes. When $R = Si(CH_3)_3$ the ht isomers fragment to a significant extent by loss of ROH and this fragmentation reaction is more pronounced for the *anti* isomer than for the *syn* isomer, which usually shows a more abundant MH[+] ion signal. Again when $R = Si(CH_3)_3$ the distinction between the hh-*syn* and hh-*anti* isomer was based on the abundance of the ion formed by the process

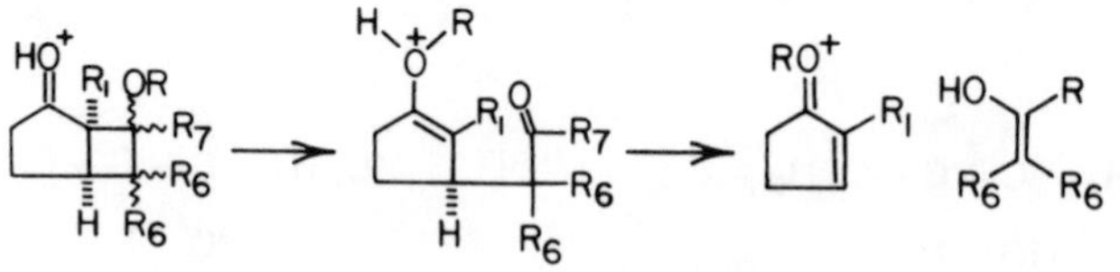

(9)

which involves migration of R and occurs more readily for the hh-*syn* isomers. When R was an acetyl group the distinction between *syn* and *anti* isomers was much less certain.

The distinction between regioisomeric hh and ht species by proton transfer CI has been achieved earlier[48] for the photodimers of Structures L and M. The fission of MH[+] to give [M/2 + H][+] is much more pronounced for the *ht* isomer than for the *hh* isomer, since in the latter β-fission forms Structure N with the charge adjacent to the carbonyl group while β-fission in the ht isomer forms the more energetically favorable ion, Structure O.

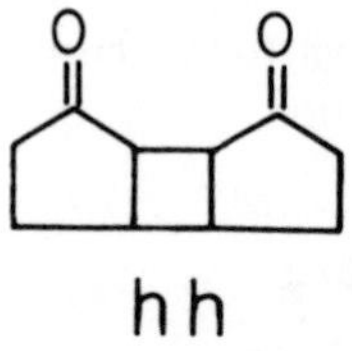

hh

STRUCTURE L

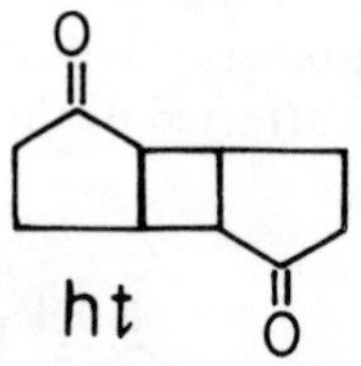

ht

STRUCTURE M

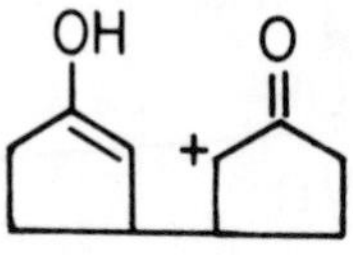

STRUCTURE N

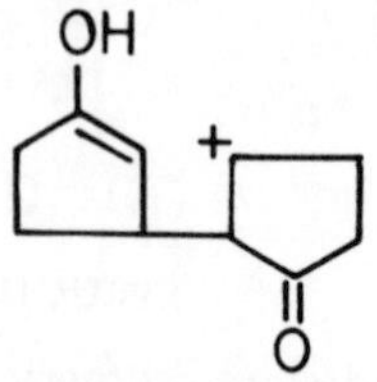

STRUCTURE O

cis-ANTI-cis

STRUCTURE P

cis-SYN-cis

STRUCTURE Q

The trimethylsiloxy (TMSO) derivatives of tricyclo[6,2,0,0$^{2.7}$]decan-3-one and tricyclo[5,2,0,0$^{2.6}$]nonan-3-one can exist in both the cis-ANTI-cis (Structure P) and cis-SYN-cis (Structure Q) configurations. In the i-C$_4$H$_{10}$ CI the MH$^+$ ions of the *SYN* configuration showed a much more pronounced elimination of (CH$_3$)$_3$SiOH while the MH$^+$ ions of the *ANTI* isomers showed a much more prominent peak originating (Reaction 10) by migration of a trimethylsilyl group to the protonated carbonyl group. The distinction between *SYN* and *ANTI* compounds is readily made.[49]

$$\text{(10)}$$

Finally, when a 1:1 mixture of diisopropyl-d$_o$-D-tartrate and diisopropyl-d$_{14}$-L-tartrate are introduced by direct insertion probe the respective MH$^+$ peaks for the labeled and unlabeled compounds are approximately equal. However the peaks at m/z 469, 483, and 497, due to the various M$_2$H$^+$ ions, are not in the expected 1:2:1 ratio: that at m/z 483 being only 46% of the expected intensity. When similar experiments were carried out with dimethyl-d$_o$-L-tartrate and dimethyl-d$_6$-D-tartrate the mixed proton-bound dimer was only 78% of the expected intensity. These results imply that the DL proton-bound dimer is slightly less stable than the DD and LL proton-bound dimers, although similar effects were not observed in mixtures of optically active camphors or N-acetylamphetamines.[50]

The results discussed in this section show that intramolecular hydrogen bonding exerts a strong influence on chemical ionization mass spectra and, therefore, provides a sensitive probe of stereochemistry. In addition, there is ample evidence that more subtle effects, not always completely understood, originate from the steric configuration of the molecule. It is clear that chemical ionization mass spectrometry is an extremely useful tool for the exploration of the stereochemistry of gas phase molecules.

IV. COLLISION-INDUCED DISSOCIATION OF IONS PRODUCED BY CHEMICAL IONIZATION

With a suitable choice of reagent gas it frequently is possible to arrange that a given compound produces largely a single ion in its chemical ionization mass spectrum. This ion may be the MH$^+$ ion in proton transfer CI, the [M−H]$^-$ ion in proton abstraction CI, or a cluster ion such as [M + Cl]$^-$. While the formation of a single product ion is advantageous for quantifying a known compound or for indicating its presence (in the absence of interfering components) in a mixture, the formation of fragment ions is essential for structural elucidation and frequently aids in identification as well. Whereas such fragment ions in many cases may be produced by selecting a more energetic ionization reaction, an alternative approach consists of recording the distribution of ionic products resulting from the collision-induced dissociation of the single ion produced in a gentle ionization reaction. The spectrum of ionic products produced in such a collision process has acquired several differing names as has the collision process itself. In the collision process the ions acquire sufficient energy to undergo fragmentation; consequently the process has been called collisional activation (CA) and the distribution of ions resulting a CA mass spectrum.[51] Alternatively, the process has been called collision-activated decomposition (CAD) and the resulting spectrum a CAD spectrum.[52] The distribution of products resulting from collision-induced dissociation (CID) often has been determined by mass analyzed ion kinetic energy (MIKE) spectrometry and the spectra have been called MIKE spectra,[53] CID/MIKE spectra,[54] or CID spectra.[55] The latter term will be used in the present work. The techniques for CID spectra have been outlined in Section VII, Chapter 3.

There are several uses of CID which have been illustrated in the literature. These range from studies designed to provide information on the structures of gaseous ions to the differentiation of isomers by the CID spectra of ionic derivatives and the identification of specific components in complex mixtures. Selected examples of these applications will be discussed in turn.

The use of CID in the structure elucidation of ions formed by electron impact was pioneered by McLafferty. Work in this area has been reviewed recently.[56,57] A number of interesting applications providing information on the structures of ions formed in CI processes have been reported by Maquestiau and co-workers.[60,61]

It has been shown[58,59] that the collision-induced dissociation of the dimeric ion $[B_1HB_2]^+$, where B_1 and B_2 are Brønsted bases bridges by a proton, leads to B_1H^+ and/or B_2H^+, with the proton preferentially associating with the base of greater proton affinity. This led Maquestiau et al.[60] to predict that if the $[M + NH_4]^+$ ion formed in the NH_3 CI of ketones is the dimeric ion $[R_1R_2C=O.H.NH_3]^+$, it should yield primarily NH_4^+ on dissociation since the proton affinity of NH_3 is greater than the proton affinities of ketones. They observed very little NH_4^+, the major dissociation reactions involving elimination of NH_3 and H_2O. They concluded that the NH_4^+ ion had added to the carbonyl bond to give a carbinolamine (Structure R) rather than forming a dimeric ion.

$$R_1 - \underset{\underset{R_2}{|}}{\overset{\overset{OH}{|}}{C}} - \overset{+}{N}H_3$$

STRUCTURE R

In a similar fashion they concluded that the $[M + NH_4 + NH_3]^+$ ion consisted of a protonated carbinolamine with hydrogen-bonded ammonia and the $[2M + NH_4]^+$ ion was a protonated carbinolamine with hydrogen-bonded ketone.

Maquestiau et al.[61] also have used CID spectra to provide information on the structure of protonated or ethylated aromatic amines found in the CH_4 CI mass spectra. For example, they found that the ions formed by $C_2H_5^+$ addition to pyrrole gave a CID spectrum which differed from the spectrum given by the ions formed by protonation of N-ethylpyrrole. They concluded that in both cases the added cation did not bond to the nitrogen, Scheme 3. From similarly designed studies they concluded that imidazole was

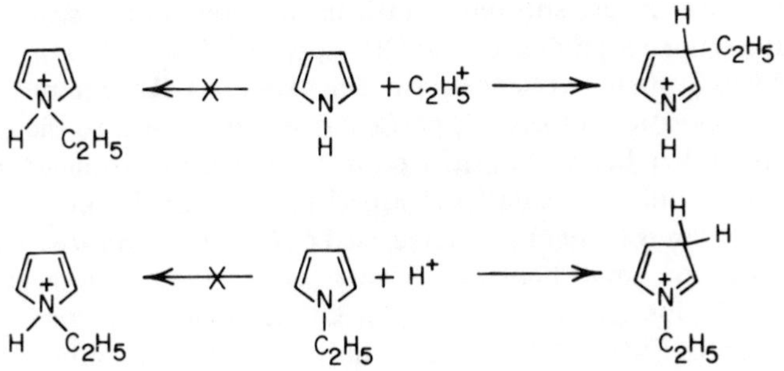

SCHEME 3

protonated or ethylated primarily on the unsubstituted nitrogen, pyridine on the nitrogen, and aniline on the aromatic ring.

In related studies Maquestiau et al.[55] have examined the structures of the $[MH^+-H_2O]$ ions formed in the proton transfer CI mass spectra of some aromatic and aliphatic oximes. They observed that the $[MH^+-H_2O]$ ion from *syn*-benzaldoxime gave a CID spectrum similar to that obtained from protonated phenyl isocyanide while the $[MH^+-H_2O]$ ion from *anti*-benzaldoxime gave a CID spectrum similar to that derived from protonated benzonitrile. They concluded that water elimination is concerted with a stereospecific migration of the substituent *anti* to the hydroxyl function, Scheme 4. Studies of other oximes were in agreement with this conclusion.

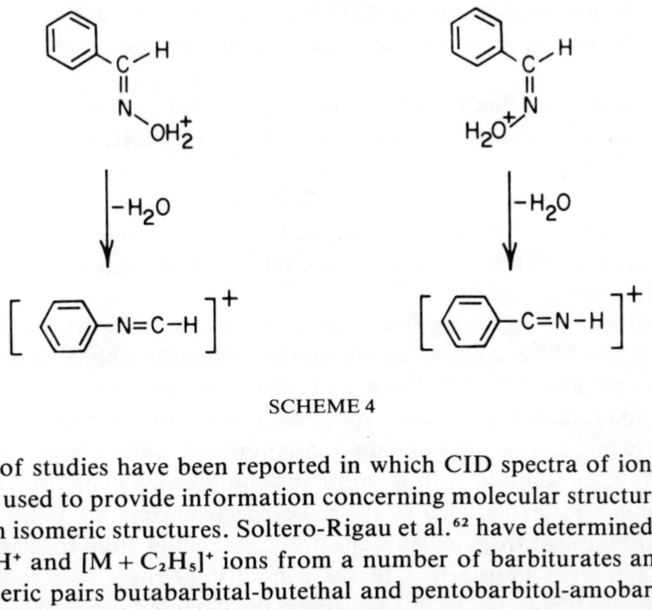

SCHEME 4

A number of studies have been reported in which CID spectra of ions produced by CI have been used to provide information concerning molecular structures or to distinguish between isomeric structures. Soltero-Rigau et al.[62] have determined the CID spectra of the MH^+ and $[M + C_2H_5]^+$ ions from a number of barbiturates and have shown that the isomeric pairs butabarbital-butethal and pentobarbitol-amobarbitol can each be distinguished. McCluskey et al.[63] have shown that *o*-hydroxybenzoic acid and *p*-hydroxybenzoic acid can be distinguished from the CID spectra of their MH^+ ions; the *para* isomer shows loss of CO_2 while the *ortho* isomer does not. In the same study it was found that the CID spectra of the $[M-H]^-$ ions derived from 3,4- and 2,5-dihydroxybenzoic acids showed sufficient differences to distinguish between the two compounds.

Sigsby et al. have reported the CID spectra for a variety of protonated ketones and ethers[64] and protonated amines and esters.[65] Distinctive and complex spectra were obtained which, in several cases, showed significant differences for structural isomers. Busch et al.[66] have reported studies of the CID spectra of the $[M-H]^-$ ions produced from polychlorinated 2-phenoxyphenols. The negative ion CID spectra showed significant differences for positional isomers, particularly with respect to the quinoxide anion formed. On the other hand the positive ion CID spectra produced as a result of charge reversal showed only very small differences for positional isomers.

Both Bradley et al.[54] and Hunt et al.[67] have used CID on CI-produced ions to provide sequence information on small peptides. Bradley et al. studied the negative ion CID spectra of the $[M-H]^-$ ions formed in the OH^- CI mass spectra of underivatized peptides. Fragmentation of $[M-H]^-$ occurred as indicated schematically in Scheme 5 for a tripeptide.

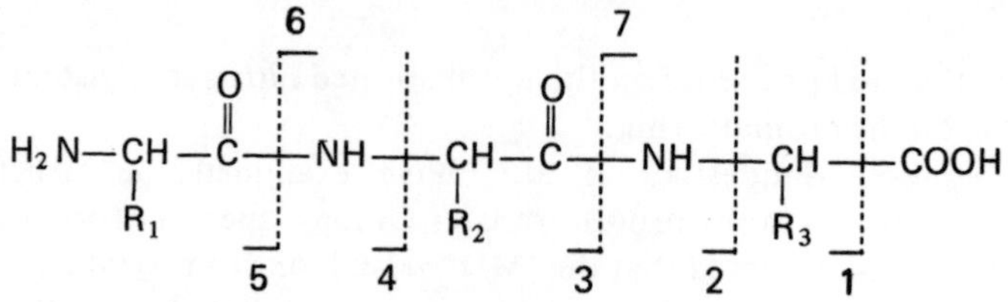

$$\underset{5}{\underset{\displaystyle H_2N-\underset{\underset{R_1}{|}}{CH}-}{\overset{6}{\overset{\displaystyle \overset{O}{\overset{\|}{C}}}{}}}\underset{4}{\Big|}NH\underset{3}{\Big|}\underset{\underset{R_2}{|}}{CH}-\overset{7}{\overset{\displaystyle \overset{O}{\overset{\|}{C}}}{}}\underset{2}{\Big|}NH\underset{1}{\Big|}\underset{\underset{R_3}{|}}{CH}\Big|COOH$$

SCHEME 5

For the tripeptides investigated, 30 to 50% of the CID signal corresponded to loss of CO_2 from $[M-H]^-$ (ion 1), a fragment which provides no structural information. Ions 2 and 3 and 4 and 5 provide complementary information. In general, ions 2, 3, 6, and 7 were of significant intensity in the CID spectra but ions 4 and 5 were of very low intensity or were absent. It is not clear how useful this approach will be for larger peptides.

Hunt et al.[67] have used the i-C_4H_{10} CI of N-acetylated-N,O-permethylated peptides to provide sequence information. As discussed in Section XIII, Chapter 5, a low intensity MH^+ ion is observed as well as two series of sequence ions, the N-terminus acyl ions (A_n^+) and the C-terminus $Z_nH_2^+$ ions (Figure 12, Chapter 5). The CID spectra of the MH^+ ions showed the characteristic A_n^+ and $Z_nH_2^+$ ions while the CID spectra of A_n^+ ions showed lower-weight A_n^+ ions and the CID spectra of ZnH_2^+ ions showed lower-weight $Z_nH_2^+$ fragments.

The potentially most useful application of collision-induced dissociations is the use of the so-called MS/MS techniques to identify individual components of complex mixtures without separation.[51,53,68,69] When a complex mixture is subjected to electron impact or chemical ionization a complex spectrum results. In many cases this complexity precludes the identification of individual components, particularly those present in minor amounts. One solution is to separate the components temporally using suitable chromatographic systems coupled to the mass spectrometer, the latter being used to identify each component from its mass spectrum as it elutes from the chromatographic system. An alternative approach is to use a multistage mass spectrometer for both separation and identification. In this approach the complex sample is ionized, preferably by soft ionization techniques such as CI which simplifies the spectrum obtained. An ion characteristic (MH^+, $[M-H]^-$, etc.) of the suspected compound is selected by the first stage of the MS/MS system and injected into a collision chamber, with the products of the collision-induced dissociation being analyzed by the second stage of the MS/MS system. In effect, the mass separation replaces the temporal separation of the chromatographic procedure while the identification by the CID spectrum replaces identification by the conventional EI or CI mass spectrum.

The component is identified by the mass of the selected ion and the, hopefully, characteristic CID spectrum which can be compared with the spectra of pure components. A less certain identification can be made by detection of a specific single fragmentation reaction. This is analogous to single-ion monitoring in chromatography/ mass spectrometry. This specific monitoring may consist of identifying either a specific fragment ion mass or identifying a specific neutral mass lost in the fragmentation process. The latter is particularly useful when classes of compounds are of interest. For example, the [M−H]⁻ ion of carboxylic acids fragment by loss of CO_2 on CID and carboxylic acids in complex mixtures can be identified by searching for those precursor ions which lose 44 amu on collision-induced decomposition.[70]

A major requirement of the MS/MS technique is that ions undergo collision-induced dissociations which are unique and characteristic of the neutral molecule from which the ions were derived. A number of studies[52,62-66,68,70,71] have concentrated largely on establishing the CID spectra of the MH^+ or [M−H]⁻ ions derived from a variety of compounds, rather than identifying components of complex mixtures. However, there have been a number of studies which have explored the utility of the technique for the detection of the components of complex mixtures. Kondrat et al.[72] used CID spectra to detect such alkaloids as cocaine, morphine, papaverine, coniine, and atropine in various plant materials. The presence of coniine in fresh stem and leaf material was established from the CID spectrum of the MH^+ ion of m/z 128 even though the m/z 128 ion intensity was not significantly above background. The stem and leaf material were ground under liquid nitrogen and introduced directly by solid probe together with ammonium acetate which served as a source of the reagent gas NH_3. In related studies[73,74] cocaine and cinnamoyl cocaine have been identified from the CID spectra of the MH^+ ions, in plant material (∼1 mg) inserted directly using a solid probe. Figure 3 shows the CID spectrum of m/z 304 obtained from cocoa leaves with the CID spectrum obtained from pure cocaine. An additional study[75] was devoted to the identification of alkaloids in crude plant extracts from the CID spectra of the MH^+ ions. Zackett et al.[76] have identified a number of components in coal liquids from the CID spectra of the MH^+ ions produced using i-C_4H_{10} reagent gas. In most cases the CID spectra were compared with those derived from authentic samples.

In negative ion studies McCluskey et al.[63] have identified benzoic acid, salicylic acid, hippuric acid, ascorbic acid, butylated hydroxyanisole (BHA), and glucose in a variety of matrixes from the CID spectra of the [M−H]⁻ ions. Both the negative ion CID spectra and the positive ion CID spectra produced by charge stripping were recorded. In the same study they also detected glucose in urine samples using the CID loss of HCl from the [M + Cl]⁻ attachment ion as a monitor. Hunt et al.,[52] using a triple quadrupole instrument, analyzed a spiked sample of sludge for p-nitrophenol, 2,4-dinitrophenol, and dioctyl phthalate. Analysis for the nitrophenols was carried out by monitoring the loss of NO and NO_2 from the [M−H]⁻ ions. Analysis of the phthalate was made by monitoring the formation of m/z 167 and m/z 149 produced (Scheme 6) on CID of the MH^+ ion.

The sensitivity of the MS/MS technique at the present time appears to be somewhat less than that of GC/MS, however, the time taken for analysis is much less. Kruger et al.[72] have reported that 10^{-11} g of p-nitrophenol can be detected by monitoring the loss of OH from MH^+. Hunt et al.[52] reported a detection limit of ∼10 ppm for nitrophenols in sludge while McCluskey et al.[63] reported a detection limit of ∼50 pg for glucose (S/N = 10). For the most part these represent ideal situations and it is not clear what the detection limit would be in real, complex samples. The amount of quantitative work which has been done is small; in principle, quantitation by the addition of isotopically labeled standards to the mixture should be possible, although matrix effects may be severe. In addition, the method undoubtedly will encounter difficulties if isomeric molecules are present in the same sample.

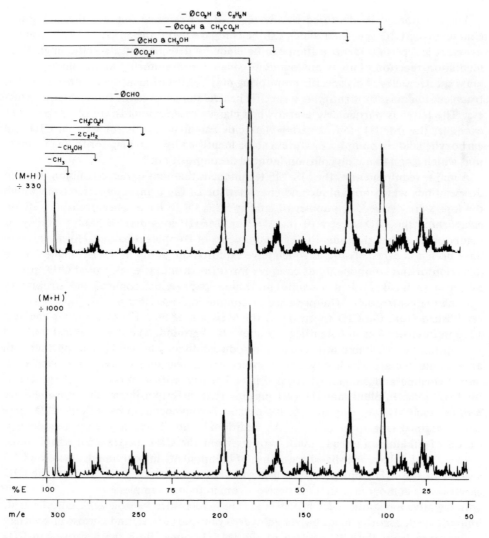

FIGURE 3.   Comparison of CID spectrum of m/z 304 from coca leaves (bottom) with the CID spectrum of m/z 304 from authentic cocaine (top). (From Kondrat, R. W. and Cooks, R. G., *Anal. Chem.*, 50, 81A, 1978. With permission.)

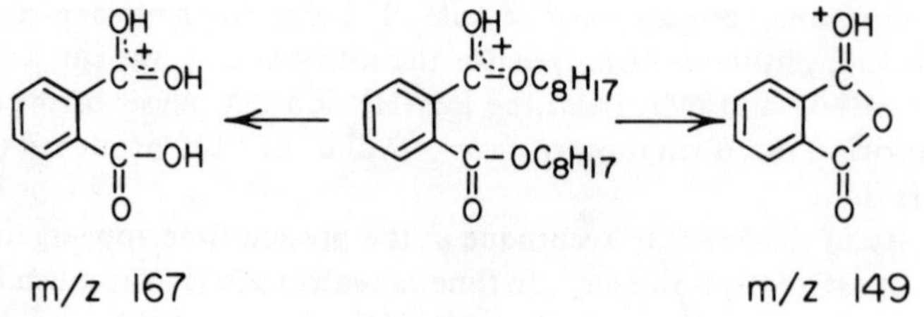

m/z 167   m/z 149

SCHEME 6

V. REACTANT ION MONITORING

When the effluent from a gas chromatograph is passed into a chemical ionization source and the intensity of the CI reactant ion is monitored one observes a decrease in the reactant ion signal as each component of the mixture elutes from the chromatograph and passes through the ion source.[77] This decrease results from consumption of the reactant ion in the chemical ionization reaction. The trace obtained by monitoring the reactant ion signal as a function of time is the equivalent of the plot of total sample ion current vs. time obtained by rapid repetitive scanning of the mass spectrometer with computer processing of the data. The technique has been called reactant ion monitoring by Hatch and Munson.[77]

Simple reactant ion monitoring offers few advantages over total sample ion current measurements and, indeed, probably is of lower sensitivity. The usefulness of the method appears to lie in the opportunity it provides to make use of the potential selectivity of the chemical ionization process. As discussed in Chapter 2, most exothermic ion/molecule reactions are fast and have roughly similar rate constants while most endothermic ion/molecule reactions are very slow. If a reagent ion can be chosen which reacts exothermically with a certain class of compounds but endothermically with others, the difference in reactivity will be reflected in the size of the "peak" in the reactant ion trace. A comparison of the reactant ion trace with the GC detector trace then permits identification of those peaks which arise from the reactive class of compounds.

In experiments designed to test this concept Hatch and Munson[77] monitored $C_4H_9^+$ (from isobutane) as a mixture of hydrocarbons and alcohols was separated in the gas chromatograph and introduced into the ion source. The $C_4H_9^+$ ion trace showed large peaks due to rapid reaction with the alcohol components but only minor peaks from the slow reaction with the hydrocarbon components of the mixture. They also showed that monitoring the $C_6H_7^+$ reactant ion produced in a $He/CH_4/C_6H_6$ mixture permitted identification of the aromatic components of an aromatic/paraffinic hydrocarbon mixture; proton transfer from $C_6H_7^+$ to paraffinic hydrocarbons is endothermic while proton transfer to aromatics is exothermic. The monitoring of $C_3H_7O^+$ (protonated acetone produced in a $He/CH_4/[CH_3]_2CO$ mixture) allowed the selective identification of the ketones present in a hydrocarbon/ketone mixture; again proton transfer from $C_3H_7O^+$ to hydrocarbons is endothermic while proton transfer to ketones is exothermic.

Subba Rao and Fenselau[78] have used selective charge exchange to identify classes of compounds in a GC trace. They showed that the mass spectrum of benzene consisted primarily of $C_6H_6^{+\cdot}$ at 0.4 torr pressure. This ion does not react readily by proton transfer and is a relatively low-energy charge transfer reagent with a recombination energy of ~ 9.2 eV.[79] Using reactant ion monitoring Subba Rao and Fenselau found that $C_6H_6^{+\cdot}$ would selectively ionize substituted benzenes with ionization energies less than 9.2 eV. They also showed that in mixtures of fatty acid esters only the unsaturated fatty acid esters were ionized by $C_6H_6^{+\cdot}$.

The technique should be applicable to negative ion studies as well, although no work in this area has been reported. The main analytical use of reactant ion monitoring appears to lie in the qualitative analysis of complex mixtures.

The area of the "peak" in the reactant ion trace is proportional to the amount of sample passing through the source and to the rate constant for reaction of the relevant ion with the sample. Thus, if the relative amounts of two components are known the relative peak areas can be used to derive relative reaction rate constants. This method has been used by Hatch and Munson[80] to measure relative rate constants for reactions of CH_5^+ and $C_2H_5^+$ with a series of hydrocarbons. The results obtained are in reasonable agreement with earlier measurements and the relative rate constants predicted by theory.

REFERENCES

1. Biemann, K., *Mass Spectrometry: Organic Chemical Applications*, McGraw-Hill, New York, 1962, chap. 5.
2. Budzikiewicz, H., Djerassi, C., and Williams, D. H., *Structure Elucidation of Natural Products by Mass Spectrometry*, Vol. 1, Holden-Day, San Francisco, 1964, chap. 2.
3. Hunt, D. F., McEwen, C. N., and Upham, R. A., Determination of active hydrogen in organic compounds by chemical ionization mass spectrometry, *Anal. Chem.*, 44, 1292, 1972.
4. Hunt, D. F., McEwen, C. N., and Upham, R. A., Chemical ionization mass spectrometry. II. Differentiation of primary, secondary and tertiary amines, *Tetrahedron Lett.*, 4539, 1971.
5. Blum, W., Schlumpf, E., Liehr, J. G., and Richter, W., On-line hydrogen/deuterium exchange in capillary gas chromatography-chemical ionization mass spectrometry (GC-CIMS) as a means of structure analysis in complex mixtures, *Tetrahedron Lett.*, 565, 1976.
6. Frieser, B. S., Woodin, R. L., and Beauchamp, J. L., Sequential deuterium exchange reactions of protonated benzenes with D_2O in the gas phase by ion cyclotron resonance spectroscopy, *J. Am. Chem. Soc.*, 97, 6895, 1975.
7. Martinson, D. P. and Buttrill, S. E., Determination of the site of protonation of substituted benzenes in water chemical ionization mass spectrometry, *Org. Mass Spectrom.*, 11, 762, 1976.
8. Stewart, J. H., Shapiro, R. H., DePuy, C. H., and Bierbaum, V. M., Hydrogen-deuterium exchange reactions of carbanions with D_2O in the gas phase, *J. Am. Chem. Soc.*, 99, 7650, 1977.
9. DePuy, C. H., Bierbaum, V. M., King, G. K., and Shapiro, R. H., Hydrogen-deuterium exchange reactions of carbanions with deuterated alcohols in the gas phase, *J. Am. Chem. Soc.*, 100, 2921, 1978.
10. Hunt, D. F. and Sethi, S. K., Gas-phase ion/molecule isotope exchange reactions: methodology for counting hydrogen atoms in specific structural environments by chemical ionization mass spectrometry, *J. Am. Chem. Soc.*, 102, 6953, 1980.
11. Farneth, W. E. and Brauman, J. I., Dynamics of proton transfer involving delocalized negative ions in the gas phase, *J. Am. Chem. Soc.*, 98, 7891, 1976.
12. Olmstead, W. N. and Brauman, J. I., Gas-phase nucleophilic displacement reactions, *J. Am. Chem. Soc.*, 99, 4219, 1977.
13. Asubiojo, O. I. and Brauman, J. I., Gas-phase nucleophilic displacement reactions of negative ions with carbonyl compounds, *J. Am. Chem. Soc.*, 101, 3715, 1979.
14. Green, M., Mass spectrometry and the stereochemistry of organic molecules, in *Topics in Stereochemistry*, Vol. 9, Allinger, N. L. and Eliel, E. L., Eds., John Wiley & Sons, New York, 1976.
15. Mandelbaum, A., Application of mass spectrometry to stereochemical problems, in *Stereochemistry, Fundamentals and Applications*, Vol. 1, Kagan, H. B., Ed., Georg Thieme Verlag, Stuttgart, 1977.
16. Green, M., Mass spectrometry — a sensitive probe of molecular geometry, *Pure Appl. Chem.*, 50, 185, 1978.
17. Morton, T. L. and Beauchamp, J. L., Chemical consequences of strong hydrogen bonding in the reactions of organic ions in the gas phase. Interactions of remote functional groups, *J. Am. Chem. Soc.*, 94, 3671, 1972.
18. Dzidic, I. and McCloskey, J. A., Influence of remote functional groups in the chemical ionization mass spectra of long chain compounds, *J. Am. Chem. Soc.*, 93, 4955, 1971.
19. Aue, D. H., Webb, H. M., and Bowers, M. T., Quantitative evaluation of intramolecular strong hydrogen bonding in the gas phase, *J. Am. Chem. Soc.*, 95, 2699, 1973.
20. Yamdagni, R. and Kebarle, P., Gas phase basicities of amines. Hydrogen bonding in proton-bound amine dimers and proton-induced cyclization of α,ω-diamines, *J. Am. Chem. Soc.*, 95, 3504, 1973.
21. Longevialle, P., Milne, G. W. A., and Fales, H. M., Chemical ionization mass spectrometry of complex molecules. XI. Stereochemical and conformational effects in the isobutane chemical ionization mass spectra of some steroidal aminoalcohols, *J. Am. Chem. Soc.*, 95, 6666, 1973.

22. Longevialle, P., Girard, J.-P., Rossi, J.-C., and Tichy, M., Influence of the interfunctional distance on the dehydration of amino alcohols in isobutane chemical ionization mass spectrometry, *Org. Mass Spectrom.*, 14, 414, 1979.

23. Longevialle, P., Girard, J.-P., Rossi, J.-C., and Tichy, M., Isobutane chemical ionization mass spectrometry of β-aminoalcohols. Evaluation of populations of conformers at equilibrium in the gas phase, *Org. Mass Spectrom.*, 15, 268, 1980.

24. Winkler, J. and McLafferty, F. W., Stereochemical effects in the chemical ionization mass spectra of cyclic diols, *Tetrahedron*, 30, 2971, 1974.

25. Munson, B., Chemical ionization mass spectrometry — analytical applications of ion-molecule reactions, in *Interactions Between Ions and Molecules*, Ausloos, P., Ed., Plenum Press, New York, 1975.

26. Van de Sande, C. C., Van Gaever, F., Hanselaer, R., and Vandewalle, M., Correlation between the stereochemistry and the chemical ionization spectra of epimeric 3,4-dimethyl-1,2-cyclopentanediols, *Z. Naturforsch.*, 32b, 810, 1977.

27. Claeys, M. and Van Haver, D., Conformation effects in the chemical ionization spectra of derivatives of isomeric 2,3-dimethyl-1,4-cyclopentanediols, *Org. Mass Spectrom.*, 12, 531, 1977.

28. Dhaenens, L., Van de Sande, C. C., and Vangaever, F., The isobutane chemical ionization mass spectra of 2-methyl substituted 1,3-cyclopentanediol diacetates and dimethyl ethers, *Org. Mass Spectrom.*, 14, 145, 1979.

29. Munson, B., Jelus, B. L., Hatch, F., Morgan, T. K., and Murray, R. K., Stereochemical effects in the mass spectra of 2-hydroxy-, 5-hydroxy-, and 2,5-dihydroxyprotoadamantanes, *Org. Mass Spectrom.*, 15, 161, 1980.

30. Respondek, J., Schwarz, H., Van Gaever, F., and Van de Sande, C. C., 1,3-Dioxolanylium und 1,3-dioxanylium-derivate via protonenkatalysierte S$_N$1 reacktionen in der gas phase, *Org. Mass Spectrom.*, 13, 618, 1978.

31. Van Gaever, F., Monstrey, J., and Van de Sande, C. C., Chemical ionization mass spectrometry of bifunctional cyclopentanes and cyclohexanes. A correlation between stereochemistry and chemical ionization spectra, *Org. Mass Spectrom.*, 12, 200, 1977.

32. Van de Sande, C. C., Van Gaever, F., Sandra, P., and Monstrey, J., Chemical ionization mass spectrometry: a useful stereochemistry probe, *Z. Naturforsch.*, 32b, 573, 1977.

33. Hunt, D. F., Reagent gases for chemical ionization mass spectrometry, *Adv. Mass Spectrom.*, 6, 517, 1974.

34. Winkler, F. J. and Stahl, D., Intramolecular ion solvation effects on gas-phase acidities and basicities. A new stereochemical probe in mass spectrometry, *J. Am. Chem. Soc.*, 101, 3685, 1979.

35. Winkler, F. J. and Stahl, D., Stereochemical effects on anion mass spectra of cyclic diols. Negative chemical ionization, collisional activation, and metastable ion spectra, *J. Am. Chem. Soc.*, 100, 6779, 1978.

36. Weinkam, R. J. and Gal, J., Effects of bifunctional interactions in the chemical ionization mass spectrometry of carboxylic acids and methyl esters, *Org. Mass Spectrom.*, 11, 188, 1976.

37. Harrison, A. G. and Kallury, R. K. M. R., Stereochemical applications of mass spectrometry. II. Chemical ionization mass spectra of isomeric dicarboxylic acids and derivatives, *Org. Mass Spectrom.*, 15, 277, 1980.

38. Benoit, F. M. and Harrison, A. G., Predictive value of proton affinity — ionization energy correlations involving oxygen-containing molecules, *J. Am. Chem. Soc.*, 99, 3980, 1977.

39. Middlemiss, N. E. and Harrison, A. G., Structure and fragmentation of gaseous protonated acids, *Can. J. Chem.*, 57, 2827, 1979.

40. Michnowicz, J. and Munson, B., Studies in chemical ionization mass spectrometry: 17-hydroxy steroids, *Org. Mass Spectrom.*, 6, 765, 1972.

41. Michnowicz, J. and Munson, B., Studies in chemical ionization mass spectrometry: steroidal ketones, *Org. Mass Spectrom.*, 8, 49, 1974.

42. Wolfshutz, R., Gransee, M., Seedorff, M., and Schwarz, H., Stereochemische effecte bei sauerkatalysierten retro-Diels-Alder reaktionen in der gasphase-eine neue analytische anwendung der chemischen ionisation, *Z. Anal. Chem.*, 295, 143, 1979.

43. Harrison, A. G. and Kallury, R. K. M. R., Stereochemical applications of mass spectrometry. I. The utility of electron impact and chemical ionization mass spectrometry in the differentiation of isomeric benzoin oximes and phenylhydrazones, *Org. Mass Spectrom.*, 15, 249, 1980.

44. Budzikiewicz, H. and Busker, E., Studies in chemical ionization mass spectrometry. III. CI-spectra of olefins, *Tetrahedron*, 36, 255, 1980.

45. Termont, D., Van Gaever, F., Dekeukeleire, D., Claeys, M., and Vandewalle, M., Differentiation between regio- and stereo-isomers of bicyclo[3,2,0]heptanone-2 derivatives by chemical ionization mass spectrometry, *Tetrahedron*, 33, 2433, 1977.

46. Claeys, M., Matveeva, H., Devreese, A., Termont, D., and Vandewalle, M., Regio and stereochemical effects in the chemical ionization mass spectra of isomeric bicyclo[3,2,0]heptan-2-one trimethylsiloxy derivatives, *Bull. Soc. Chim. Belg.*, 87, 375, 1978.

47. Claeys, M., Van Audenhove, M., and Vandewalle, M., Evidence for the occurrence of an oxygen to acyl shift in 7-acetoxybicyclo[3,2,0]heptan-2-ones upon chemical ionization mass spectrometry, *Bull. Soc. Chim. Belg.*, 88, 799, 1979.

48. Ziffer, H., Fales, H. M., Milne, G. W. A., and Field, F. H., Chemical ionization mass spectrometry of complex molecules. III. The structures of the photodimers of α,β-unsaturated ketones, *J. Am. Chem. Soc.*, 92, 1597, 1970.

49. Van Audenhove, M., Dekeukeleire, D., and Vandewalle, M., Mass spectrometry used to discriminate between cis-SYN-cis and cis-ANTI-cis photo adducts, *Bull. Soc. Chim. Belg.*, 89, 371, 1980.

50. Fales, H. M. and Wright, G. J., Detection of chirality with the chemical ionization mass spectrometer. "Meso" ions in the gas phase, *J. Am. Chem. Soc.*, 99, 2339, 1977.

51. McLafferty, F. W., Tandem mass spectrometry (MS/MS): a promising new technique for specific component determination in complex mixtures, *Accts. Chem. Res.*, 13, 33, 1980.

52. Hunt, D. F., Shabanowitz, J., and Giordani, A. B., Collision activated decomposition of negative ions in mixture analysis with a triple quadrupole mass spectrometer, *Anal. Chem.*, 52, 386, 1980.

53. Kondrat, R. W. and Cooks, R. G., Direct analysis of mixtures by mass spectrometry, *Anal. Chem.*, 50, 81A, 1978.

54. Bradley, C. V., Howe, I., and Beynon, J. H., Sequence analysis of underivatized peptides by negative ion chemical ionization and collision induced dissociation, *Biomed. Mass Spectrom.*, 8, 85, 1981.

55. Maquestiau, A., Van Haverbeke, Y., Flammang, R., and Meyrant, P., Chemical ionization mass spectrometry of some aromatic and aliphatic oximes, *Org. Mass Spectrom.*, 15, 80, 1980.

56. McLafferty, F. W., Structure of gas phase ions from collisional activation spectra, in *Chemical Applications of High Performance Mass Spectrometry*, Gross, M. L., Ed., American Chemical Society, Washington, D.C., 1977.

57. Schlunegger, U. P., *Advanced Mass Spectrometry*, Pergamon Press, Oxford, 1980.

58. Cooks, R. G. and Kruger, T. L., Intrinsic basicity determination using metastable ions, *J. Am. Chem. Soc.*, 99, 1279, 1977.

59. McLuckey, S. A., Cameron, D., and Cooks, R. G., Proton affinities from dissociation of proton bound dimers, *J. Am. Chem. Soc.*, 103, 1313, 1981.

60. Maquestiau, A., Flammang, R., and Nielsen, L., A study of some cations formed in the ammonia chemical ionization of ketones using mass analyzed ion kinetic energy spectrometry, *Org. Mass Spectrom.*, 15, 376, 1980.

61. Maquestiau, A., Van Haverbeke, Y., Mispreuve, H., Flammang, R., Harris, J. A., Howe, I., and Beynon, J. H., The gas phase structure of some protonated and ethylated aromatic amines, *Org. Mass Spectrom.*, 15, 144, 1980.

62. Soltero-Rigau, E., Kruger, T. L., and Cooks, R. G., Identification of barbiturates by chemical ionization and mass analyzed ion kinetic energy spectrometry, *Anal. Chem.*, 49, 435, 1977.

63. McCluskey, G. A., Kondrat, R. W., and Cooks, R. G., Direct mixture analysis by mass-analyzed ion kinetic energy spectrometry using negative chemical ionization, *J. Am. Chem. Soc.*, 100, 6045, 1978.

64. Sigsby, M. L., Day, R. J., and Cooks, R. G., Fragmentation of even electron ions. Protonated ketones and ethers, *Org. Mass Spectrom.*, 14, 273, 1979.

65. Sigsby, M. L., Day, R. J., and Cooks, R. G., Fragmentation of even electron ions. Protonated amines and esters, *Org. Mass Spectrom.*, 14, 556, 1979.

66. Busch, K. L., Norström, A., Nilsson, C. Å., Bursey, M. M., and Hass, J. R., Negative ion mass spectra of some polychlorinated 2-phenoxyphenols, *Environ. Health Perspect.*, 36, 125, 1980.

67. Hunt, D. F., Buko, A. M., Ballard, J. M., Shabanowitz, J., and Giordani, A. B., Sequence analysis of polypeptides by collision activated dissociation on a triple quadrupole mass spectrometer, *Biomed. Mass Spectrom.*, 8, 397, 1981.

68. McLafferty, F. W. and Bockhoff, F. M., Separation/identification system for complex mixtures using mass separation and mass spectral characterization, *Anal. Chem.*, 50, 69, 1978.

69. Cooks, R. G., Mixture analysis by mass spectrometry, in Trace Organic Analysis: A New Frontier in Analytical Chemistry, NBS Spec. Publ. 519, Hertz, H. S. and Chester, S. N., Eds., U.S. Department of Commerce, Washington, D.C., 1979.

70. Zackett, D., Schoen, A. E., Kondrat, R. W., and Cooks, R. G., Selected fragment scans of mass spectrometers in direct mixture analysis, *J. Am. Chem. Soc.*, 101, 6781, 1979.

71. Kruger, T. L., Litton, J. F., Kondrat, R. W., and Cooks, R. G., Mixture analysis by mass-analyzed ion kinetic energy spectrometry, *Anal. Chem.*, 48, 2113, 1976.

72. Kondrat, R. W., Cooks, R. G., and McLaughlin, J. L., Alkaloids in whole plant material: direct analysis by kinetic energy spectrometry, *Science*, 199, 978, 1978.

73. Kondrat, R. W., McCluskey, G. A., and Cooks, R. G., Multiple reaction monitoring in mass spectrometry/mass spectrometry for direct analysis of complex mixtures, *Anal. Chem.*, 50, 2017, 1978.

74. Youssefi, M., Cooks, R. G., and McLaughlin, J. L., Mapping of cocaine and cinnamoylcocaine in whole coca plant tissues by MIKES, *J. Am. Chem. Soc.*, 101, 3400, 1979.

75. Kruger, T. L., Cooks, R. G., McLaughlin, J. L., and Ranieri, R. L., Identification of alkaloids in crude extracts by mass analyzed ion kinetic energy spectrometry, *J. Org. Chem.*, 42, 4161, 1977.
76. Zackett, D., Shaddock, V. M., and Cooks, R. G., Analysis of coal liquids by mass-analyzed ion kinetic energy spectrometry, *Anal. Chem.*, 51, 1849, 1979.
77. Hatch, F. and Munson, B., Reactant ion monitoring for selective detection in gas chromatography/chemical ionization mass spectrometry, *Anal. Chem.*, 49, 731, 1977.
78. Subba Rao, S. C. and Fenselau, C., Evaluation of benzene as a charge exchange reagent, *Anal. Chem.*, 50, 511, 1978.
79. Lindholm, E., Mass spectra and appearance potentials studied by use of charge exchange in a tandem mass spectrometer, in *Ion-Molecule Reactions,* Franklin, J. L., Ed., Plenum Press, New York, 1972.
80. Hatch, F. and Munson, B., Relative rate constants for reactions of CH_5^+ and $C_2H_5^+$ with hydrocarbons by gas chromatography-chemical ionization mass spectrometry, *J. Phys. Chem.*, 82, 2362, 1978.

INDEX